PUBLIC HEALTH IN THE 21ST CENTURY

# *TOXOPLASMA GONDII*

# PREVALENCE IN HUMANS AND ANIMALS, GENETIC STRUCTURE AND ROLE IN DISEASE DISTRIBUTION

# PUBLIC HEALTH IN THE 21ST CENTURY

Additional books in this series can be found on Nova's website under the Series tab.

Additional E-books in this series can be found on Nova's website under the E-book tab.

PUBLIC HEALTH IN THE 21ST CENTURY

# *TOXOPLASMA GONDII*

# PREVALENCE IN HUMANS AND ANIMALS, GENETIC STRUCTURE AND ROLE IN DISEASE DISTRIBUTION

MELVIN B. GOODSON
AND
DON WINSLOW
EDITORS

New York

**Library of Congress Cataloging-in-Publication Data**

ISBN: 978-1-62618-349-0
Library of Congress Control Number: 2013933834

*Published by Nova Science Publishers, Inc. † New York*

# Contents

# Preface

*Toxoplasma gondii* is a parasite of the phylum Apicomplexa, Coccidia order, being an obligate intracellular protozoan that affects humans and diverse vertebrate hosts. In this book, the authors discuss its prevalence in humans and animals, genetic structure and role in disease distribution. Topics include the congenital transmission of toxoplasma gondii in ovine; the role of chickens in the transmission of toxoplasmosis to humans; updates on the serological and molecular diagnosis of toxoplasmosis; epidemiology and control of toxoplasmosis in dogs, cats, bovine, and in public health.

Chapter I – Toxoplasmosis is for the most part of individuals an asymptomatic infection. Pregnant women, their fetus and immunocompromised patients require a differentiated clinical and laboratory approach because, when they are infected, they may progress severely or result in death. The diagnostic methods can be indirect, detecting antibodies in serum, or direct indicating the parasite or its DNA by molecular methods (polymerase chain reaction, PCR and its variants). Serological diagnosis is indicated in pregnant women in early pregnancy, newborns of infected mothers; patients with chorioretinitis; donors and transplant recipients, patients with lymphadenopathy and fever. The serological tests which are most commonly used are ELISA, ELFA or ISAGA that detect antibodies of the class IgA, IgM, IgE and IgG. The antibodies are related to the time of infection - the immunoglobulins of the class IgA, IgM and IgE are produced in the first week (acute phase). The IgM decreases around six months, but may be detectable for years. IgA can also be found for months. Thus, IgM and IgA should not be considered markers of recent infection. Sabin Feldman tests, immunofluorescence or agglutination test detect earlier antibodies of the IgG class over other techniques (ELISA). IgG antibodies are detected one week after infection, decreasing from three months, but with persistence of residual titer throughout life. The need to estimate the time of occurrence of the infection, such as cases of pregnant women who have IgM and probable acute infection, the IgG avidity test is advisable. The infection is considered recent (three months or less) when the percentage of antibodies with low avidity is <30%, and correlates with increased risk of fetal infection. The diagnosis is made by the newborn detection of IgM and / or IgA and / or the presence of IgG in bonds rising or persistently positive at 12 months of a child's life. The IgM and IgA do not exceed passively the placenta and are considered markers of congenital infection. However, the IgG antibodies that cross the placenta are found in a significant number of uninfected children. In these cases, the diagnosis of infection in newborn or infant is accomplished by monitoring the IgG titers or the differentiation of maternal IgG to the newborn, primarily through the enzymatic immunofiltration assays (ELIFA) and Western blot (immunoblot - IB IgM and IgG ). The analysis of IFN-Î³ release was recently described as a complementary method in the diagnosis of congenital toxoplasmosis. Despite advances in research on *Toxoplasma gondii* in the last 20 years, the serological diagnosis of toxoplasmosis remains difficult, since the high sensitivity of current techniques to detect the presence of residual IgM titers for years, turning its interpretation complicated.

The direct detection of the parasite by molecular techniques such as PCR and Real Time PCR, can aid the interpretation of the actual state of human / parasite interaction. The PCR has special value in the diagnosis and therapeutic management if positive in amniotic fluid, blood of the fetus or newborn, cerebrospinal fluid, brain tissue or placenta.

Chapter II – *Toxoplasma gondii* is a member of the phylum Apicomplexa, order Coccidia, and is an obligate intracellular protozoan parasite that infects humans and several vertebrate hosts. In its life cycle, T. gondii has three infectious forms: the sporozoites (oocysts), tachyzoites (individually or in groups) and bradyzoites (tissue cysts). Toxoplasmosis is a zoonosis of major spread worldwide. Sheep toxoplasmosis was first described by Olafson and Monlux in the United States.These authors reported this disease in a female who presented neurological symptoms, hyperthermia, muscle rigidity, and other clinical signs. The diagnosis was made after necropsy and histopathological examination of the brain and spinal cord of the animal. In small ruminants, toxoplasmosis is relatively common, responsible for reproductive problems, generating great economic losses in sheep flocks, where the infection is the main cause of abortions, fetal malformations, premature births and neonatal deaths being the main route of infection the ingestion of sporulated oocysts of the parasite.

Chapter III – Considering the close proximity between men and dogs and the zoonotic potential of these diseases, determining the occurrence of endoparasitosis, mainly toxoplasmosis, in pets has become increasingly relevant. This disease is frequently diagnosed in spite of the existent therapeutic and prophylactic measures. Studies of animal toxoplasmosis are essential in view of its transmission to men and its pathogenicity in pets. The dog can be involved in the mechanical transmission of the infection by *Toxoplasma gondii* and in spite of not being the definitive host, it has an epidemiological role in this disease. The high rate of dogs naturally infected by *T. gondii* and the great possibility of adquiring this protozoan, justify the importance of attempting against for the occurrence of this illness. The aim of this chapter will be to clear some clinical aspects, address the participation of small animals in the spread of the disease, as well as discuss methods of diagnosis, therapeutic measures, prophylaxis and control of this disease.

Chapter IV – *Toxoplasma gondii* is a parasite of the phylum *Apicomplexa*, *Coccidia* order, being an obligate intracellular protozoan that affects humans and diverse vertebrate hosts. The genus was named *Toxoplasma* (from the greek *toxon* = arc, *plasma* = shape) referring to its shape of "first quarter" or "half moon" and *gondii* in reference to the rodent where the parasite was isolated. In retrospect, the correct name for the parasite should have been *Toxoplasma gundii*, however the host was incorrectly identified as *Ctenodactylus gondi*. The felines, in particular the domestic cats, are the definitive hosts of the parasite, whereas warm-blooded animals are the intermediate hosts. Toxoplasmosis is one of the most spreaded zoonoses in the whole world, however the frequency of infection started to be characterized from 1948, and then with other researchers, developing the first diagnostic of the disease. The toxoplasmosis affects about two million people worldwide. The prevalence of seropositive individuals for toxoplasmosis increases with age and differs depending on the population cultural patterns, dietary habits and urban or rural origin. It is known that in areas of tropical or subtropical humid climate, the prevalence is higher, once this type of weather favors the survival of oocysts in the environment. It is estimated that about 70% of the brazilian population has already been infected at some point in life. In Brazil, it was found that in every 1,000 live births, five children presented the disease. In different countries, the seroprevalence has been shown between 0% to 90% in the human population. Some countries as Thailand and Japan present low prevalence (<20%). However, Australia, Poland, UK and Belgium show average prevalence (between 23 and 53%), while Brazil, Tahiti and France present high prevalence (> 60%). The importance of the animal toxoplasmosis occurs, at first, by the fact that infected animals serve as direct or indirect source of the human infection and, secondly, by the various reproductive changes such as abortion, neonatal mortality and congenital defects consequent to the *T. gondii* infection, which represent significant losses in animals of economic interest and company. In the world literature on detection of anti-*T. gondii* in bovines, discrepant values are found, ranging from 0 to 99%. In Brazil, this variation in occurrence values is also observed ranging from 1.03% to 71.00%. Infection rates given for sheep flocks in Brazil, vary from 8% to 55%. The infection of herbivores occurs primarily through ingestion of oocysts present in food and contaminated soils. Infection of herds may be associated with the ingestion of grains contaminated by oocysts and the stabling of animals at sites containing feline manure may increase the risk of infection. Direct contact with felines had little epidemic consequence, although the presence of those infected animals indicates a contaminated environment, incurring risk to the human population and other animals.

Chapter V – Toxoplasmosis is an anthropozoonosis caused by the protozoan *Toxoplasma gondii* (*T. gondii*), an obligate intracellular parasite belonging to the phylum Apicomplexa, Sporozoea Class, Coccidia Sub-class, Eucoccidiida Order, Eimeriina Sub-order, Sarcocystidae Family and Toxoplasmatinae Sub-Family. It is a protozoan capable of infecting nucleated cells from a wide variety of mammals and birds as well as primary cell cultures of fish and insects. It presents a cosmopolitan geographic distribution, with high rates of infection in humans. It is estimated that one third of the world population has already been exposed to the *T. gondii*. The first description of the species was carried out by Alfonso Splendore in Brazil in 1908, who reported the presence of the protozoan parasitizing laboratory rabbits. In that same year, the French parasitologists Charles Nicolle and Louis Hebert Manceaux isolated it in an African rodent (*Ctenodactylus gondii*), at the Pasteur Institute of Tunisia. Originally denominated *Leishmania gondii*, by its similarity with the protozoan of the genus *Leishmania*, the adequacy of its nomenclature only occurred in the following year. The life cycle of the *T. gondii* was only elucidated more than 50 years after the first report of its discovery. In the late 1960s, the analysis of cat fecal material made it possible to discover an evolutive type of protozoan, which was capable of inducing infection in other mammals and birds by its ingestion. Its coccidial cycle was elucidated in the 1970s, when the sexual stages of *T. gondii* were found in small intestine of cats. Although the protozoan *T. gondii* is considered the only valid species in its genus, three clonal genetic lineages were initially recognized. These are known as Type I, Type II and Type III, and they are predominant in Western Europe and the United States. However, surveys on samples of this parasite from other regions of the world, such as South American localities, have revealed other genotypes, differing from the archetypes mentioned above. When several markers were applied against Brazilian strains, a highly reticulated phylogenetic structure was identified, which suggests that recombination plays an important role in strain diversification in South America. Currently, it is known that its life cycle is divided between feline and non-feline infections, which are correlated with sexual and asexual replication, respectively. Definitive hosts are members of the family Felidae, for example domestic cats.

Asexual stages of toxoplasma-like parasites were first observed at the turn of the century in tissues of non-felines, for example birds and mammals. They are probably all warm-blooded animals including most livestock, and humans. In intermediate hosts, *T. gondii* undergoes two phases of asexual development. In the first phase, tachyzoites (or endozoites) multiply rapidly by repeated endodyogeny in many different types of host cells. Tachyzoites of the last generation initiate the second phase of development which results in the formation of tissue cysts. Within the tissue cyst, bradyzoites (or cystozoites) multiply slowly by endodyogeny. Tissue cysts have a high affinity for neural and muscular tissues. They are located predominantly in the central nervous system (CNS), the eye as well as skeletal and cardiac muscles. Unsporulated oocysts are released into the intestinal lumen and passed into the environment with the faeces. Sporogony occurs outside the host and leads to the development of infectious oocysts which contain two sporocysts, each containing four sporozoites. Several studies have been carried out on the asexual stages of *T. gondii*, in particular on the tachyzoite, while much fewer studies have considered the sexual stages or their infectious product, i.e. the sporozoites within the oocyst. It is noteworthy that cats and wild felids are essential to the persistence of *T. gondii* in hosts such as grazing animals (e.g. sheep, deer and, more recently, chickens) because they serve as the sole source of the infectious oocysts that contaminate the environment. Environmental contamination by *T. gondii* oocysts is a concern from both public health and biodiversity perspectives. Surface run-off containing feline feces containing *T. gondii* oocysts enters in multiple ecosystems. Currently ingesting undercooked meat also provides an important avenue for *T. gondii* infection through the ingestion of viable tissue cysts. Undercooked pork is another common risk factor for *T. gondii* exposure; beef is not believed to be a major player, however, mammals and avians raw meat was found to be a risk factor in several epidemiological studies in developed and underdeveloped countries.

Even as other important risk factors in evidence are for infection include not washing kitchen knives after they have been used to cut raw meat, fruits, and vegetables, as well as infrequent hand washing. In general, the infection by the *T. gondii* has a higher prevalence in tropical or subtropical most climate zones, which favors the viability of oocysts in the environment. The prevalence of reagents for anti-*T. gondii* antibodies increases with age and it varies depending on the cultural patterns of the population, its hygiene, food and urban and rural provenance. The prevalence of the disease in children is relatively low, increasing with age and the time of exposure to risk factors throughout life. In Brazil, the official statistics from the Ministry of Health indicate toxoplasmosis as responsible for 90 deaths annually. However, there is a lack of data on morbidity and mortality caused by this parasite, so this data is formed by sub-estimated numbers.

*Toxoplasma Gondii* strain involved in the infection and host's immune status, the disease may cause severe illness in human beings and several other classes of animals. Over the last decades, toxoplasmosis has stimulated the interest of many researchers, mainly by public health relevance and by pervasive affected systems in the hosts, either as an opportunistic disease in immunocompromised patients and pregnant women, or as adaptative disease in immunocompetent animals where reproductive and neural systems is most affected. Millions of people worldwide have a serological prevalence of between 15% and 85% in different regions of the planet. In the United States, the prevalence of seropositivity ranges from 10% to 30%, approximately 14% in the reproductive age. In Norway, for example, the positivity in serologic tests is only 10.9%. On the other hand, the prevalence is 28% in Denmark and it reaches 36.7% in Austria. In Malaysia, a recent study found a rate of prevalence of 49%, reaching rates of up to 70% in France, especially in Paris. In Brazil, it is estimated that 70% of the population will manifest clinical signs of toxoplasmosis at some point in life. Surveys conducted in pregnant women have shown high seropositivity rates, ranging from 40% to 80%. The frequency of seropositive pregnant women for IgG anti-Toxoplasma varies widely among regions, from 28.1% to 70.9% in the South, 70.7% to 98.8% in the Midwest, 50.1% to 90.0% in the Southeast and 66.4% to 72.7% in the Northern Region of the country. In the Northeast, information about the frequency of infections in pregnant women, capable of perinatal transmission, is scarce; in the Sergipe State there was a prevalence rate of 69% in a study performed with 9.550 pregnant women, in Salvador city the rate was of 69.3%, 71.5% in Ceará, 74.7% in Recife, in João Pessoa it was observed a rate of 91% of prevalence in a study performed with 55 pregnant women, 86.2% in

Alagoas, 66.38% in São Luis and in Natal a rate of prevalence of 66% in a study comprising 190 pregnant women was observed. In the North of Country, in Belém, a rate of 73% was obtained in a study conducted with 531 pregnant women, a rate of 76.2% in Amapá and reaching levels above 70% in Amazonas city. In the Southeast, in the capital city of the state of São Paulo the prevalence rate of 67.7% was observed in a study comprising 481 pregnant women and, in Rio de Janeiro, 77.1%; 73.5% in Espírito Santo and 51.6% in Uberlândia. In the South region, in the capital city of Rio Grande do Sul, a prevalence rate of 61.1% was obtained by evaluating 10.468 pregnant women and the rate of 49.2% in Paraná. In the central region of the country, in Goiania, the prevalence rate has remained stable over the past 23 years. Nearly all species of animals are susceptible to infection by the protozoan, the seropositivity rates and clinical signs of toxoplasmosis in different hosts are varied.

Chapter VI – Toxoplasmosis, a zoonosis of worldwide distribution, has importance in human and veterinary medicine. Animals can be direct or indirect source of infection to man. The infection can be acquired by ingestion of tissue cysts in infected meat, ingestion of food or water contaminated with oocysts and congenitally, by tachyzoites crossing the placenta from the infected mother to the fetus. Cats may act as intermediate or definitive hosts of this parasite. Among the infected animals, felines have an important role in the epidemiology of this agent because like definitive hosts eliminate oocysts in the environment. Evaluations of the occurrence of *Toxoplasma gondii* antibodies in felines has drawn the attention of research worldwide. The aim of this chapter will be clear some clinical manifestations aspects, address the true participation of cats in the spread of the disease, as well as discuss methods of diagnosis, therapeutic measures, prophylaxis and control of this disease.

Chapter VII – Infection by *Toxoplasma gondii* is particularly interesting from the point of view of public health. Attention should be drawn to the risk factors for acquiring this prenatal infection, considering its pathogenesis and sequelae. Women in initial stages of gestation may undergo miscarriage, premature birth, neonatal death and even the classical Sabin's Triad, characterized by retinochoroiditis, cerebral calcifications, hydrocephaly or microcephaly. The risk of acquiring this disease is higher in the postnatal life, when severe behavioral changes such as attention deficit and schizophrenia may occur, reducing the life quality of individuals. There is strong association of reactivation of this disease in cases of acquired immunodeficiency syndrome – AIDS, commonly leading to secondary infection of the central nervous system in immunocompromised people and severe encephalitis. The aim of this chapter wil be discuss educational programs directed at reducing environmental contamination by *T. gondii* would eventually decrease the cost of treating humans with clinical toxoplasmosis.

Chapter VIII – Toxoplasmosis is a zoonotic disease caused by coccidia *Toxoplasma gondii,* that affects almost all homeothermic animal species, including birds, mammals and also humans. The implementation of control programs including preventive measures such as laboratory tests frequently promotes early diagnosis and appropriate indication of adequate antiparasitic treatment to pregnant women, and is, therefore, able to diminish the seriousness of the sequelae of toxoplasmosis in the fetus. Important aspects of the control of congenital toxoplasmosis programs for maternal and neonatal screening, the toxoplasmosis control program were reviewed.

In: *Toxoplasma Gondii* ISBN: 978-1-62618-349-0
Editors: M. Goodson and D. Winslow 

*Chapter I*

# Updates on Serological and Molecular Diagnosis of Toxoplasmosis

*Marcia Antunes Fernandes and Aline Cardoso Caseca Volotão*
Fluminense Federal University, Niteroi, Rio de Janeiro, Brazil

## Abstract

Toxoplasmosis is for the most part of individuals an asymptomatic infection. Pregnant women, their fetus and immunocompromised patients require a differentiated clinical and laboratory approach because, when they are infected, they may progress severely or result in death. The diagnostic methods can be indirect, detecting antibodies in serum, or direct indicating the parasite or its DNA by molecular methods (polymerase chain reaction, PCR and its variants). Serological diagnosis is indicated in pregnant women in early pregnancy, newborns of infected mothers; patients with chorioretinitis; donors and transplant recipients, patients with lymphadenopathy and fever. The serological tests which are most commonly used are ELISA, ELFA or ISAGA that detect antibodies of the class IgA, IgM, IgE and IgG. The antibodies are related to the time ofÂ infection - the immunoglobulins of the class IgA, IgM and IgE are produced in the first week (acute phase). The IgM decreases around six months, but may be detectable for years. IgA can also be found for months. Thus, IgM and IgA should not be considered markers of recent infection. Sabin Feldman tests, immunofluorescence or agglutination test detect earlier antibodies of the IgG class over other techniques (ELISA). IgG antibodies are detected one week after infection, decreasing from three months, but with persistence of residual titer throughout life.

The need to estimate the time of occurrence of the infection, such as cases of pregnant women who have IgM and probable acute infection, the IgG avidity test is advisable. The infection is considered recent (three months or less) when the percentage of antibodies with low avidity is <30%, and correlates with increased risk of fetal infection. The diagnosis is made by the newborn detection of IgM and / or IgA and / or the presence of IgG in bonds rising or persistently positive at 12 months of a child's life. The IgM and IgA do not exceed passively the placenta and are considered markers of congenital infection. However, the IgG antibodies that cross the placenta, are found in a significant number of uninfected children. In these cases, the diagnosis of infection in newborn or infant is accomplished by monitoring the IgG titers or the differentiation of maternal IgG to the newborn, primarily through the enzymatic immunofiltration assays (ELIFA) and Western blot (immunoblot - IB IgM and IgG ). The analysis of IFN-Î³ release was recently described as a complementary method in the diagnosis of congenital toxoplasmosis. Despite advances in research on *Toxoplasma gondii* in the last 20 years, the serological diagnosis of toxoplasmosis remains difficult, since the high sensitivity of current techniques to detect the presence of residual IgM titers for years, turning its interpretation complicated. The direct detection of the parasite by molecular techniques such as PCR and Real Time PCR, can aid the interpretation of the actual state of human / parasite interaction. The PCR has special value in the diagnosis and therapeutic management if positive in amniotic fluid, blood of the fetus or newborn, cerebrospinal fluid, brain tissue or placenta.

# Introduction

The diagnosis of the infection caused by *Toxoplasma gondii* can be done in a direct manner isolating the parasite through the bioassay, cell culture, molecular diagnosis or immunehistochemistry using PCR (Polymerase Chain Reaction) in blood, amniotic fluid or placenta. The direct identification of the parasite by cell culture or isolation through bioassay is considered gold standard for the definition of the infection, however, they are slow, laborious and costly methods, apart from present low sensibility. It is known that parasitemia is detectable intermittently in some patients. Serology is the one of the major methods which is used to establish the diagnoses of toxoplasmosis an indirect way, based on the research of the antibodies against the parasite.

The laboratory is usually asked to contribute to the diagnosis of the toxoplasmosis. Often, serological diagnosis is the only way available for elucidation of the studied case. However, since the infection in most individuals is asymptomatic, serological diagnosis is usually retrospective, being indicated only in situations where it is necessary to determine the immune status of the individual, like: a) pregnant women, especially in the first trimester of pregnancy; b) patients with uveitis or chorioretinitis that does not have the history of congenital infection, c) donors and organ recipients. Besides these indications, there are patients with fever or lymphadenopathy that need of serological diagnosis, which aims to differentiate from other infections such as cytomegalovirus (CMV), Epstein-Barr virus or HIV, as well as non-infectious diseases such as hematologic malignancies.The serological diagnosis is based not only on the demonstration of specific antibodies but also in their serum level.

The results from a laboratory may be different when compared with tests carried out in other laboratories. The differences can be explained by the use of tests with different sensitivities, due to technical problems as kits of the reagents, the type of equipment or how the reactions were executed. It is clear that the standardization of all these factors is virtually impossible, but it is important to use the appropriate serum reference, standardized. The serological results have only probability value, and should be interpreted with caution, because the immune response to infection with *Toxoplasma gondii* may have individual variations. Moreover, the high sensitivity of some serological tests that can detect acute phase antibodies for a long period after infection, can lead to misinterpretation.

## A. Serology Response to the Infection Caused by *Toxoplasma Gondii*

The antibodies of the type IgA and IgM are produced during the first week and reach a plateau during the first month after infection. The IgE immunoglobulins are also detected early, but they decrease quickly. The levels of specific IgM antibodies usually decrease after one to six months, disappearing in 25% of individuals in less than seven months.

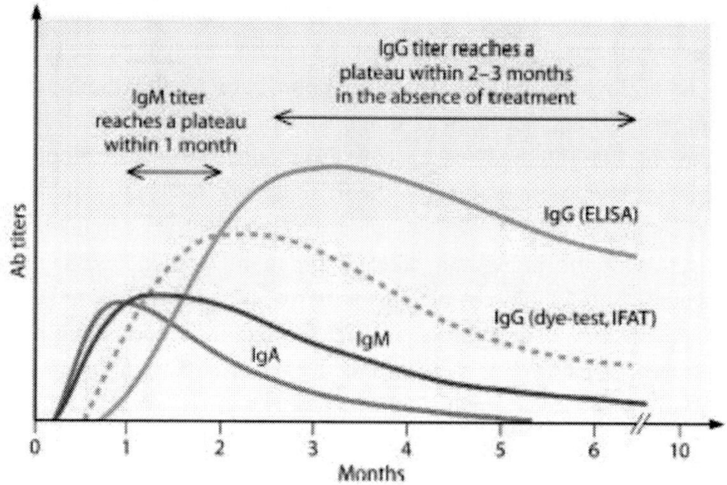


Clinical Microbiology Reviews. April 2012 Volume 25 Number 2.

Figure 1. Kinetics of the antibody (Ab) response. The average kineties of the different isotypes are represented, but they may differ among patients and accoding to the serologic technique used. IgM can be detected for years after infection.

The early detection of IgG antibodies depends on the technique used. Serological methods which use membrane antigens or whole parasite like Sabin and Feldman test (dye test), or immunofluorescence agglutination test identifies specific IgG more precocious. The technique Enzyme Linked Fluorescent Assay (ELISA) which uses mixtures of cytosolic antigens, metabolic or surface depending on the manufacturer identifies the serum IgG later. The variability in time of detection of specific IgG therefore depends on the technique used as well as individual variations. IgG is identified in serum one to three weeks after the initial increase of IgM. The IgG synthesis reaches a plateau within two to three months after infection, regardless of the technique used. After, the levels of IgG continue to decrease until they reach the residual titles, which persist throughout life.

## B. Major Serologic Tests

Several serological tests have been developed over the years, in order to facilitate the diagnosis and to identify the period of occurrence of the infection by Toxoplasma gondii. They are based on antibodies of different classes of immunoglobulins (IgG, IgM, IgA, and, more rarely, IgE) anti-*T. gondii* and according to the serological profile presented by the individual it is possible to determine the stage of infection (acute or chronic). Serological tests which are the most commonly used are Enzyme-Linked Immunosorbent Assay (ELISA) conventional catch (IgM or IgA) or IgG avidity for the Agglutination Immunosorbent Assay (ISAGA) and immunoblotting (IB).

### *B.1. Sabin and Feldman (Dye Test)*

The dye test was described in 1948 by Sabin and Feldman. It is considered even today as the gold standard test because of its high sensitivity and specificity, although few laboratories still done it. Their titles are detectable within one to two weeks after acquiring the infection. The test is based on the facility that the parasites have of staining with methylene blue. The toxoplasmas used in the reaction are alive, and if there is a specific antibody, they will be lysed by the antigen-antibody reaction in the presence of a complement.

### *B.2. Test Immunofluorescence (IF)*

It has the advantage of being realized with preserved toxoplasmas, fixed on microscope slides, making it very practical. It identifies IgM or IgG antibodies by the utilization of specific conjugates. The main disadvantage is the IF frequency of false-positive and false-negative. The false-positive results of the IF-IgM can occur by interference of rheumatoid factor, if present in the serum, even in newborns. The false-negative results of the IF-IgM can occur by the presence of IgG antibodies that prevents IgM antibodies fix to parasite antigens. Currently, most laboratories use the technique Enzyme Linked Fluorescent Assay (ELISA) to detect routine specific IgG or IgM.

### *B.3. Immunoenzymatic Assays: Enzyme-Linked Immuno-Sorbent Assay (ELISA) and Enzyme-Linked Fluorescent Assay (ELFA)*

The ELISA assay is based on a reaction in which the enzyme is bound to its substrate hydrolyzed, being added an anti-human immunoglobulin. The product is capable of developing color (or fluorescence), whose intensity is read on a spectrophotometer (or fluorometer). The amount of antibody in the serum is directly proportional to color intensity (ELISA) or fluorescence (ELFA). The ELISA can detect IgM, IgA, IgE, IgG and, being widely available in most laboratories.

These tests eliminate false results because antibodies use serum containing "anti-globulins" which we want to capture. The final product of these reactions can be measured by the color (double-sandwich ELISA), fluorescence (ELFA) or by agglutination (ISAGA). The determination of IgM, IgA, IgE and IgG can be performed by ELISA, ELFA and ISAGA.

### *B.4. IgG Avidity Test*

This test was developed to help to set the time when infection occurred. The avidity corresponds to the strength of binding between the antigen and antibody, which increases progressively over time (functional affinity). Low avidity antibodies are produced in the early stages of infection, dating three months or less. The antibodies of high avidity are more characteristic of chronic infections, or more than four months. The increase in affinity of IgG results from a process that involves the selection of antigen and B cells that direct and increase the affinity of the antigen-antibody binding site. The IgG avidity to *Toxoplasma gondii* can be performed by ELISA (enzyme-linked immunosorbent assay).

The method was originally developed in Finland and is available as a commercial kit in Europe. Avidity tests are ordered to help if the infection occurred among the last four to five months. Thus, it is very helpful when asked by pregnant women in the first trimester of pregnancy who have positive titers for IgM and IgG anti-*Toxoplasma gondii*. If a pregnant woman has the avidity test with high avidity antibodies in the first trimester of pregnancy, she did not present acute infection in the last three months.

When the infection was acquired before pregnancy the fetus is not at risk. There are reports of congenital transmission when the woman is infected or weeks before a period very close to the design, but they are exceptional situations.

The interpretation of the results of IgG avidity should be careful, since there are situations where women have positive titers for IgM anti-Toxoplasma antibodies and low avidity but no acute infection.

Some women have low avidity antibodies that persist for more than a year. Moreover, a substantial number of patients have boderlines or equivocal results. The most appropriate decision on how to proceed in treating these patients with low avidity results or mistakes should be made based on the results of other serological methods. With this objective, a series of serological tests are usually done in reference laboratories. The ELISA serological tests consist of IgM, IgA and IgE test, Sabin and Feldman and differential agglutination test. Thus, the avidity test should not be used alone as a definitive test for decision making.

### *B.5. Western Blot IgG and IgM*

Serological diagnosis of congenital toxoplasmosis is, in the majority of cases, made by demonstration of IgM and IgA anti-*Toxoplasma gondii* in newborns, since they do not usually cross the placental barrier. However, extreme caution is recommended when interpreting the presence of IgM and IgA in newborns, since it can occur maternal transmission of these antibodies during childbirth, in exceptional situations.

The average life of these antibodies transferred to the fetus during labor is low, and it is indicated to repeat the IgM in two to four days and IgA in ten days. Moreover, some newborns with congenital toxoplasmosis may not present IgM and IgA or both, during the first days of life. In normal conditions, it is not advisable to perform the diagnosis of congenital toxoplasmosis in newborns by the demonstration of antibodies of the IgG class, since IgG is passively transferred from mother to fetus. In order to explain whether the IgG present in the serum of these newborns and infants is by maternal transfer or by immune response of the infants to the infection, it is recommended that monthly tests should be conducted on infants to detect IgG titers for six months or more.

Maternal antibodies decrease in approximately half of the initial titles in each month of life. The infant who persists with IgG anti-Toxoplasma gondii to a year of life is considered with confirmed congenital infection. During the study of new methods for diagnosis of the fetus and newborn (NB), it was demonstrated the usefulness of Western blot test, pairing the serum of mothers and their newborns.

Pinon and Gruson in 1982 demonstrated that antibodies formed in response to the fetus congenital infection recognize different antigens of *Toxoplasma gondii* than those of the mother. Hofflin and Remington (1985) had identified different antigens by antibodies of the IgG and IgM produced by the mother and her child congenitally infected. Hofflin and Remington (1985) have also documented that the reaction bands by Western blot containing IgG and / or IgM serum of congenitally infected infants were not present in the serum of their mothers.

Since the first description of the use of the reaction of the Western blot by Remington et al. in 1985, ten years have passed for this technique return to be mentioned in the medical literature. Noteworthy, the abstracts published in the International Conference held Toxoplasmosis in Denmark in June 2003, indicating the use of the Western blot for diagnosis of toxoplasmosis. Recently, Tissot-Dupont et. al. published prospective studies of routine use of the Western blot IgA, IgM, IgG for diagnosis of congenital toxoplasmosis.

Western blot was more sensitive than IgM immunosorbent agglutination assay (ISAGA) (86.9 versus 69.6%) and specificity of 96.1% and 92.2%, respectively. The sensitivity increases when both techniques are applied. The use of IgA Western blot is less sensitive than the combination IgM and IgG.

When the serum obtained in the first days of life reveals a negative result, the Western blot should be interpreted with caution, since the result can become positive weeks after birth. There are situations described where the IgM is negative at birth, but the Western blot is positive in the first month of life. Further, the prenatal treatment during pregnancy, as well as the treatment of postnatal infant may produce false negative results. When the immunoblot band intensity of the infant is greater than the bands identical to her mother, the result should be considered equivocal, since this pattern has been observed in uninfected infants.

Western blot pairs from mother and her newborn seems to be very useful when not found IgM and / or IgA in infants by conventional methods, whose mothers have a definitive diagnosis of acute infection or with a high degree of suspicion during pregnancy. Thus, it is important to emphasize that the methods should be used in combination with other serological tests like ELISA or IgM and IgA ISAGA. In addition, recognizing that no test isolated or in combination with other serological tests, including Western blot allows the diagnosis of all cases.

### *B.6. Gama Interferon (Whole – Blood Gamma Interferon Release Assay)*

Cellular immunity plays a major role in resistance to infection with Toxoplasma gondii, but is not routinely evaluated for the diagnosis of infection. The secretion of interferon-γ (IFN-γ) is increased in patients infected acutely or chronically. It is observed an increased secretion of IFN-γ occurs when there is a stimulus with parasite antigens. The test is simple and is based on the increase of IFN-γ after stimulation with parasite antigens.

It is only used one ml of heparinized blood which is centrifuged, the plasma is reserved for routine serology centrifuged and the remaining material is placed in a culture medium for 24h. Subsequently, toxoplasma antigens are added, cells are centrifuged and the supernatant is analyzed for the research of IFN-γ. This test showed a sensitivity and specificity of 94% and 98%, respectively.

This way, it can be a good diagnostic method for congenital toxoplasmosis in the newborn. When the serological results are negative, this test can exclude the diagnosis of congenital toxoplasmosis at birth, thus avoiding unnecessary anxiety and "follow-up" with serial serological tests. Furthermore, a positive result aid in the decision to initiate treatment early, preventing further sequels. This test also proved to be useful in the diagnosis of perinatal tuberculosis.

The use of purified antigens or synthetic peptides with the intention of improving the performance of the test is still being studied. This test can be a useful tool for screening possible antigens with potential use in vaccines in humans. However, its implementation requires standardization routine, since there are not commercially available tests.

## C. Neonatal Screening

Neonatal screening for the identification of IgM anti-*Toxoplasma gondii* in capillary dry spot blood of newborns (NB), absorbed into filter paper cards, is carried out in some countries like USA (Massachusetts) since 1986, Denmark since 1999 and most recently in Poland. The positive dried blood should be confirmed in serum, opportunity that should also be done to investigate maternal serology.

As it uses the structure of screening programs for other diseases (phenylketonuria, hypothyroidism, etc.), it is considered a low cost alternative practice. The sensitivity of IgM in dried blood for diagnosis of congenital infection appears to be equal to the IgM serum. There were dissenting opinions concerning neonatal screening, with the claim that the IgM (antibody chosen for screening) was not present in 100% of infected newborns and remained positive for a short period of time leading to false negatives. However, several studies, such as the one of Guerina et al, a Danish study, and other study made in Brazil, point out the neonatal screening as a valid strategy, identifying congenital toxoplasmosis cases, which were unidentified in prenatal screening. They indicate that even in countries where prenatal care for identification of toxoplasmosis during pregnancy is not standardized, newborn screening is appropriate and should be introduced universally.

## D. Molecular Tests

Molecular techniques have been used for the diagnosis of congenital toxoplasmosis, ocular and in immunocompromised patients. The amplification of specific segments of the parasite DNA by the polymerase chain reaction (PCR) allows the identification of *T. gondii* in clinical samples, even when present in small numbers. Molecular diagnosis is most commonly performed on amniotic fluid, placenta, umbilical cord blood of the fetus, in cerebrospinal fluid, brain biopsy.

Initially, the target most used was the single-copy of P30 gene which encodes SAG1, the major surface antigen of T. gondii. Literature data have shown that conventional PCR for detection of single copy genes, such as the P30 gene, are less sensitive.

In 1987 Boothroyd et al. described the B1 gene, which is one of the most used. The B1 gene has 2.2 kb and repetitive nature in the genome of T. gondii, which has 35 copies of this gene. Several initiators were described to amplify various regions of this gene. Studies have demonstrated that the PCR of this gene has high sensitivity and is capable to detect even a parasite about 100,000 host cells. However, recently it was demonstrated the low specificity of this method since it was shown to co-amplification of target sequences in other human chromosomes. Nevertheless, the gene B1 still remains the most widely used in diagnosis target parasitemia.

Other target sequences has also been used for the molecular diagnosis of toxoplasmosis, such as: 1) encoding gene of the smallest subunit of ribosomal RNA that is present in 110 copies in the genome, 2) fragment of 529 base pairs, called AF146527 element, with 200 to 300 copies, 3) genes encoding α and β tubulin, both single-copy and 4) TGR1E, a segment of repetitive non-coding DNA.

The diagnosis of toxoplasmosis by the PCR is not yet fully standardized and there is still no consensus on the conditions of the best method and of the target sequence in the parasite, which directly affects the sensitivity and specificity of the reaction.

Recent studies of population structure of *T. gondii* showed the existence of variant strains intraspecific, known as types. In South America, *T. gondii* is genetically and biologically divergent, while in North America and Europe presents clonal population structure with three main lineages, types I, II and III. Type I is the most virulent when compared with Types II and III. The ones isolates from humans and other animals in South America are genetically different from those in North America and Europe, and severe toxoplasmosis in immunocompetent humans is associated with an atypical genotype characteristic of this continent. Literature data suggests that *T. gondii* spreads in a clonal way in Europe, Africa and North America, while in the strains circulating in South America occurs sexual recombination. More studies on the molecular epidemiology of toxoplasmosis involving serologic and molecular genotyping should be developed to better understand the epidemiological settings and the dynamics of the parasite transmission as well as the distribution of strains of *T. gondii.*

Most laboratories use the sequence that is repeated 35 times in the B1 gene, although some laboratories in Europe are exchanging it for the sequence of AF146527, repeated 200 to 300 times in the genome of *T. gondii.* Yet others suggest the use of the 18S ribosomal DNA of the parasite. The three techniques showed no difference in sensitivity and specificity, no significant differences occurring from a statistical viewpoint. The completion of PCR as diagnostic for toxoplasmosis becomes a good option because it is a noninvasive technique depending on the type of sample, rapid shown high specificity and sensitivity.

Guy and Joynson have demonstrated the importance of molecular diagnosis complementing and confirming the serological results, especially in cases of acute infection. PCR is often requested to perform prenatal diagnosis of congenital infection using amniotic fluid.

The amniotic fluid PCR for detection of Toxoplasma gondii-specific DNA made from the 18th week of gestation is more sensitive, faster and safer than conventional methods that used the umbilical cord blood of the fetus (cordocentesis).

The prenatal diagnosis of congenital toxoplasmosis is performed by monitoring through ultrasound images and through the PCR of amniotic fluid. The PCR test for amniotic fluid is indicated in all pregnant women with positive serological tests for acute or very suggestive. Likewise, it is indicated when there is evidence of involvement by fetal ultrasonography (hydrocephalus and / or calcifications).

French studies like Hohlfield et al. found a positive predictive value (PPV) of PCR in amniotic fluid of 100% and a negative predictive value (NPV) of 99.7%, while in others such as Romand et al., the PPV was 100% and NPV of 88%. In the study of Romand et al. the sensitivity was 64%, but they assessed the sensitivity and NPV by using a serological monitoring of newborns and infants until a definitive diagnosis during the first year of life.

Thus, as expected, the diagnosis of all cases of transmission of toxoplasmosis during gestation were not able to be performed during the prenatal period.

However, in the study of Hohlfeld et al. there was no follow-up evaluation of these children in the first year of life, being justified the variation in sensitivity and NPV. The sensitivity of prenatal diagnosis of PCR is significantly higher when maternal infection occurs between 17 weeks and 21 weeks, compared with the infection occurred before 17 weeks or after 21 weeks of pregnancy (p <0.02).

Children born from mothers chronically infected with the parasite, but that are immunocompromised by infection (eg. AIDS) or use of immunosuppressants should perform PCR of cerebrospinal fluid, blood and urine tests to exclude the congenital toxoplasmosis. In pregnant women infected with HIV-1, it is not advisable to use the PCR of amniotic fluid, as this procedure can contaminate their fetus. The examination of placental tissue is a complementary form for the diagnosis of congenital infection at birth, in addition to serological screening of newborns. The demonstration of the parasite in placental tissue can be confirmed by inoculation into mice or by the DNA of *Toxoplasma gondii* through PCR. Several retrospective studies examining placentas, showed 42-71% sensitivity and 92-100% specificity, depending on the technique used. The combination of both techniques achieves a sensitivity of 74%.

The Real-Time PCR is a technique recently introduced for diagnosis of toxoplasmosis. It combines the amplification phase and the detection of PCR products in a single step, making it faster. It reduces the usual time of the PCR result of 24 to 48h to 4h. An additional advantage of this method is the possibility to estimate the concentration of parasites in the amniotic fluid, which may be useful in determining the prognosis of RN. In the study of Romand et al. was documented that maternal infections acquired before the 20th week, with a parasite load greater than 100/ml of the amniotic fluid, related to the most severe clinical forms in infected children.

The diagnosis of chorioretinitis should be confirmed and monitored by a specialist (ophthalmologist). In cases in which is observed a positive serological result to *Toxoplasma gondii* and the ophthalmologic examination revealed lesions characteristic of toxoplasmosis it is recommended to start treatment.

However, there are patients who do not show clinical response to treatment and who have atypical lesions, making diagnosis difficult. In these cases it is recommended to perform the laboratorial diagnostic methods, such as local searching antibodies in ocular fluids or detection of the parasite by PCR in the aqueous humor or the vitreous fluid. The sensitivity of PCR in the aqueous humor or the vitreous fluid varies from 16 to 55%. Some studies have shown that the DNA of the parasite is most often detected in the aqueous humor of immunocompromised patients, but there are few reports, which prevent us to draw a definite conclusion.

## E. Transplantation: Handling of Donor and Recipient

The realization of the serologic profile of toxoplasmosis in all patients to be transplanted varies depending on countries, based primarily in prevalence of infection in each region.

In most cases, the risk of infection occurs when the donor is positive and the recipient's is negative (D + / R-), especially in transplants of heart, heart-lung, kidney, liver and pancreas. In those patients who had organ transplants and are on immunosuppressive therapy, the possibility of reactivation of chronic infection is high, which may progress to severe clinical picture, and may lead to death due to myocarditis and systemic disease. Patients who do not have immunity must be accompanied in the post-transplant, due to the risk of contracting the infection during this period. These data suggest the importance of establishing the serological profile of toxoplasmosis before transplantation, which are applied to appropriate medical procedures in these patients after this procedure.

The cornea donor who is seropositive for toxoplasmosis can perform the transplantation, although the receiver needs to be warned about the possibility of reactivation after transplantation. About 50% of seronegative recipients who receive heart of toxoplasmosis seropositive donors, develop primary acute infection, unless it receives specific prophylaxis.

The onset of clinical infection may develop in the first seven days after transplantation. However, most occurs around the second month, being indicated the chemoprophylaxis specific for toxoplasmosis in order to avoid symptomatic infection. In countries with a low seroprevalence of toxoplasmosis, cotrimoxazole used for prevention of *Pneumocystis* pneumonia also appears to offer some protection against the risk of primary infection by T. gondii. There is no consensus on the duration of chemoprophylaxis, but it is recommended not to exceed six months. After discontinuation of prophylaxis, some transplants centers recommend perform a serologic screening every six months in order to detect seroconversion or reactivation.

It has been recommended to use TMP-SMX (160 mg-800 mg) daily or three times per week for prophylaxis of toxoplasmosis in transplanted patients who require prophylaxis also *Pneumocystis* pneumonia. In the case of intolerance to TMP-SMX, pyrimethamine is used alone (25mg/day) alternatively to the prophylaxis of toxoplasmosis, whereas aerosolized pentamidine can be used to protect *Pneumocystis* pneumonia.

Patients under treatment with aerosol pentamidine as prophylaxis for *Pneumocystis* pneumonia, are not protected against infection with *Toxoplasma gondii.* Therefore, it is recommended to start the secondary prophylaxis as soon as there are severe clinical manifestations with both therapies: TMP-SMX (30) or pyrimethamine and sulfadiazine (25 mg 2g-2g-50mg daily or three times per week).

# Conclusion

There was a great improvement in the last 20 years in the research field of toxoplasmosis. Stand out the development of new serological methods and molecular techniques which have aided greatly the diagnosis of infection, especially in patients in whom accurate diagnosis, medical management changes and its clinical outcomes. The isolation of genotypes of strains of various species, on different continents, allowed to recognize that the species are not identical throughout the world and, also, vary in their virulence. Meanwhile, the development of other diagnostic markers is still required to assist the clinician faced with difficult cases, especially those with ocular and central nervous system.

# References


Bacal F, Souza-Neto JD, Fiorelli AI, Mejia J, Marcondes-Braga FG, Mangini S, et al. II Diretriz Brasileira de Transplante Cardíaco. *Arq. Bras. Cardiol.,* 94(1 supl.1):e16-e73, 2009.

Bastien, P. Diagnosis. Molecular diagnosis of toxoplasmosis. *Trans R. Soc. Trop. Med. Hyg.,* 96( supl. 1): 205-15, 2002.

Buchbinder,S ,Blatz,R, Rodloff, A. C. Comparison of real-time PCR detection methods for B1 and P30 genes of Toxoplasma gondii. *Diagn. Microbiol. Infect. Dis., 45*: 269-71, 2003.

Burg, J. L., Grover CM, Pouletty, P., Boothroyd, J.C. Direct and sensitive detection of a pathogenic protozoan, Toxoplasma gondii, by polymerase chain reaction. *J. Clin. Microbiol.,* 27(8):1787-92, 1989.

Beazley, D M, Egerman, R.S. Toxoplasmosis. *Semin. Perinatol.,* 22(4): 332-338,aug, 1998.

Bèssieres M H, Berrebi A, Rolland M, Bloom MC, Roques C et al. Neonatal screening for congenital toxoplasmosis in a cohort of 165 women infected during pregnancy and influence of in utero treatment on the results of neonatal tests. *Eur. J. Obstet. Gynecol. Reprod. Biol.*, 94: 37-45, 2001.

Burg JL, Grove CM, Pouletty P and Boothroyd JC. Direct and sensitive detection of a pathogenic protozoan, *Toxoplasma gondii,* by polymerase chain reaction. *J. Clin. Microb.*, 27: 1787-1792, 1989.

Camargo, M E Diagnóstico de Laboratório da Toxoplasmose humana. .21(1): 3-11,1989.

Camargo,M E. Toxoplasmose. In:Ferreira, A Walter, àvila, Sandra L.M. Diagnóstico laboratorial das principais doenças infecciosas e auto-imunes. Rio de Janeiro: *Guanabara. Koogan.*, 302.Cap19:165-173, 1996.

Carellos E V M , de Andrade G M Q , Aguiar R AL P. Evaluation of prenatal screening for toxoplasmosis in Belo Horizonte, Minas Gerais State, Brazil: a cross-sectional study of postpartum women in two maternity hospitals.*Cad.Saúde Pública,* Rio de Janeiro, 24(2):391-401,fev 2008.

Connell T, N, Bar-Zeev and Curtis N. Early detection of perinatal tuberculosis using a hole blood interferon-gamma release assay. *Clin. Infect. Dis.*, 42:e82–e85, 2006.

Costa, JM, Pautas C, Ernault P, Foulet F, Cordonnier C and S. Bretagne. Real-time PCR for diagnosis and follow-up of toxoplasma reactivation after allogeneic stem cell transplantation using fluorescence resonance energy transfer hybridization probes. *J. Clin. Microbiol.*, 38: 2929–2932, 2000.

Costa, JM, Pautas C, Ernault P, Foulet F, Cordonnier C and S. Bretagne. Real-time PCR for diagnosis and follow-up of toxoplasma reactivation after allogeneic stem cell transplantation using fluorescence resonance energy transfer hybridization probes. *J. Clin. Microbiol.*, 38: 2929–2932, 2000.

Costa, J M, Ernault P, Gautier E, and Bretagne S. Prenatal diagnosis of congenital toxoplasmosis by duplex real-time PCR using fluorescence resonance energy transfer hybridization probes. *Prenat. Diagn.*, 21:85–88, 2001.

Derouin F, Pelloux H. Prevention of toxoplasmosis in transplant patients. *Clin. Microbiol. Infect.*, 14:1089 –1101, 2008.

Di Cristina, M, Del Porto P, BuffolanoW, Beghetto E, Spadoni A, Guglietta S, Piccolella E, Felici F and Gargano N. The Toxoplasma gondii bradyzoite antigens BAG1 and MAG1 induce early humoral and cell-mediated immune responses upon human infection. *Microbes Infect.*, 6:164–171, 2004.

Fardeau C, Romand S, Rao NA, Cassoux N, Bettembourg, Thuliez P, Lehonang P. Diagnosis of toxoplasmic retinochoroiditis with atypical clinical features. *Am. J. Ophthalmol.*, 134:196 –203, 2002.

Fekkar A, Ajzenberg D, Bodaghi B, Touafek F, Hoang P Le et al. Comparison of immunoblotting, calculation of the Goldmann-Witmer coefficient, and real-time PCR using aqueous humor samples for diagnosis of ocular toxoplasmosis. *J. Clin. Microbiol.,* 46:1965–1967, 2008.

Fernandez-Sabe N. Risk factors, clinical features, and outcomes of toxoplasmosis in solid-organ transplant recipients: a matched case-control study. *Clin. Infect. Dis.,* 54:355–361, 2011.

Filice GA, Hitt JA, Mitchell CD, Blackstad M , Sorensen SW. Diagnosis of *Toxoplasma* parasitemia in patients with AIDS by gene detection after amplification with polymerase chain reaction. *J. Clin. Microb.,* 31: 2327-2331, 1993.

Florence Robert- Gangneux and Marie-Laure Dardè. Clinical Microbiology Reviews: Epidemiology of and Diagnosis Strategies for toxoplasmosis. April, 25( 2): 265-289, 2012.

Foudrinier F, Villena I, Jaussaud R, Aubert D, Chemla C et al. Clinical value of specific immunoglobulin E detection by enzyme-linked immunosorbent assay in cases of acquired and congenital toxoplasmosis. *J. Clin. Microbiol.,* 41:1681–1686, 2003.

Fricker-Hidalgo H, Brenier-Pinchat,MP, Schaal, JP, Equy,V, Bost-Bru,C et al. Value of Toxoplasma gondii detection in one hundred thirty-three placentas for the diagnosis of congenital toxoplasmosis. *Pediatr. Infect. Dis. J.,* 26: 845– 846, 2007.

Fricker-Hidalgo H, Pelloux, H, Racinet, C, Grenstette,I, Bost-Bru,C et al. 1998. Detection of Toxoplasma gondii in 94 placentae from infected women by polymerase chain reaction, in vivo, and in vitro cultures. *Placenta,* 19:545–549, 1998.

Fuentes, I, M, Rodriguez, C J Domingo, F Del Castillo, Juncosa T and Alvar J. Urine sample used for congenital toxoplasmosis diagnosis by PCR. *J. Clin. Microbiol.,* 34:2368–2371, 1996.

Garweg JG, Jacquier P, Boehnke M. Early aqueous humor analysis in patients with human ocular toxoplasmosis. *J. Clin. Microbiol.,* 38:996–1001, 2000.

Gras L, Gilbert RE, Wallon M, Peyron F, Cortina-Borja M. Duration of the IgM response in women acquiring Toxoplasma gondii during pregnancy: implications for clinical practice and cross-sectional incidence studies. *Clin. Microl. Biol.,* .132:541–548, 2004.

Greig JR, Holliman R. Neonatal screening for toxoplasma infection. *Lancet,* 354:1030-1, 1999.

Grob, U, Roggenkamp, A, Janitschke, K, Heesemann, J. Improved sensitivity of de polymerase chain reaction for detection of Toxoplasma gondii in biological and human clinical specimens. *Eur. J. Clin. Microbiol. Infect. Dis.,* 11(1): 33-9, 1992.

Guay, J-M., Dubois, D., Morency, M.J., Gagnon, S., Mercier, J., Levesque, R.C. Detection of the pathogenic parasite Toxoplasma gondii by specific amplification of ribosomal sequences using comultiplex polymerase chain reaction. *J. Clin. Microbiol.,* 31( 2):203-7, 1993.

Guerina NG, Hsu HW, Meissner HC, Maguire JH, Lynfield R, Stechenberg B, et al. Neonatal serologic screening and early treatment for congenital Toxoplasma gondii infection.The New England Regional Toxoplasma Working Group. *N. Engl. J. Med.,* 330(26):1858-63, 1994.

Guy EC, Jooynson DH. Potencial of the polymerase reaction in the diagnosis of active *Toxoplasma* infection by detection of parasite in blood. *J. Infect. Dis.,* 172(1):319-22, 1995.

Hedman,K, Lappalainen M, Seppala I, and O. Makela. 1989. Recent primary Toxoplasma infection indicated by a low avidity of specific IgG. *J. Infect. Dis.,* 159:736–739, 1989.

Hodgall, E. Vuust, J., Lind, P., Petersen, E. Characterization of Toxoplasma gondii isolates using polymerase chain reaction (PCR) and restriction fragment length polymorphism (RFLP) of the non-coding Toxoplasma gondii (TGR)-gene sequences. *Int. J. Parasitol.,* 30:853-8, 2000.

Homan, W. L., Vercammen, M., De Braekeleer, J., Verschueren, H. Identification of a 200- to 300-fold repetitive 529bp DNA fragment in Toxoplasma gondii, and its use for diagnostic and quantitative PCR. *Int. J. Parasitol.,* 30: 69-75, 2000.

Hofflin JM, Remington JS. Tissue culture isolation of *Toxoplasma* from blood of a patient with AIDS. *Arch. Inter. Ed.,* 145: 925-926, 1985.

Hohlfeld, P, F, Daffos, J, Costa M, Thuylliez P, Forestier F and Vidaud M. Prenatal diagnosis of congenital toxoplasmosis with polymerase-chain-reaction test on amniotic fluid. *N. Engl. J. Med.,* 331:695–699, 1994.

Kompalic-Cristo A, Nogueira SA, Guedes Al, Frota C, Gonzalez LF, Brandão A et al. Lack of technical specificity in the molecular diagnosis of toxoplasmosis. *Trans. R. Soc. Trop. Med. Hyg.,* 98:92-95, 2004.

Lago E.G. Estratégias de controle da toxoplasmose congênita. Porto Alegre; Pontifícia Universidade Católica do Rio Grande do Sul; 2006.

Lamoril, J, Molina, JM, de Gouvello, A, Garin, YJ, Deybach, JC, Modaï, J, Derouin, F Detection by PCR of Toxoplasma gondii in blood in the diagnosis of cerebral toxoplasmosis in patients with AIDS. *J. Clin. Pathol.,* 49:89-92, 1996.

Lappallainen, M, KosKiniem, M, Hillesmaa, V, Ammala P., Teramo K. et al. Outcome of children after maternal primary Toxoplasma infection during pregnancy with emphasis on avidity of specific IgG. *Pediatr. Infect. Dis.,* J.14: 354-361, 1995.

Lebech,M, Joynson DHM, Seitz HM, Thulliez P ,Gilbert RE, Dutton GN, Ovlisen B, Petersen E. Classification system and case definitions of Toxoplasma gondii infection in immunocompetent pregnant women and their congenitally infected offspring. *Eur. J. Clin. Microbiol. Infect. Dis.,* 15:799-805, 1996.

Lebech,M, Joynson DHM, Seitz HM, Thulliez P ,Gilbert RE, Dutton GN, Ovlisen B, Petersen E. Classification system and case definitions of Toxoplasma gondii infection in immunocompetent pregnant women and their congenitally infected offspring. *Eur. J. Clin. Microbiol. Infect. Dis.,* 15:799-805, 1996.

Lebech M, Anderson O, Christensen NC, Hertel J, Nielsen HE, Petersen B et al. Feasibility of neonatal screening for *Toxoplasma* infection in the absence of prenatal treatment. Danish Congenital Toxoplasmosis Group. *Lancet,* 353(9167):1834-7, 1999.

Lynfield, R., Guerina, N.G. Toxoplasmosis. *Pediatr. Rev.,* 18(3): 75-85, mar, 1997.

McLeod, R, Boyer K, Karrison T, Kasza K, Swisher C, Roizen N, Jalbrzikowski J, Remington J, Heydemann P, Noble AG, Mets M, Holfels E, Withers S, Latkany P, and P. Meier. Outcome of treatment for congenital toxoplasmosis, 1981–2004: the National Collaborative Chicago-Based, Congenital Toxoplasmosis Study. *Clin. Infect. Dis.,* 42:1383–1394, 2006.

Montoya,J.G. and J. S. Remington.Studies on the serodiagnosis of toxoplasmic lymphadenitis. *Clin. Infect. Dis.,* 20:781–790, 1995.

Montoya JG, Parmley S, Liesenfeld O, Jaffe GJ, Remington JS. Use of the polymerase chain reaction for diagnosis of ocular toxoplasmosis. *Ophthalmology,* 106:1554-1563, 1999.

Montoya JG, Huffman H B, and emington JS, Abstract.40$^{th}$ Annu. *Meet. Infect. Dis. Soc. Am. abstr.,* 270:.92, 2002.

Montoya JG, Liesenfeld O. Toxoplasmosis. Abstr. 40th Annu. *Meet. Infect. Dis. Soc. Lancet,* 363:1965–1976, 2004.

Nascimento FS, Suzuki LA, Rossi CL. Assessment of the value of detecting specific IgA antibodies for the diagnosis of a recently acquired primary Toxoplasma infection. *Prenat. Diagn.,* 28:749 –752,2008.

Oliveira, JD, Aguiar ME Pinto, JFC. Toxoplasmose congênita: aspectos epidemiológicos e laboratoriais. *Clin. Pediátr. Set.,*32-43, 1986.

Paul M, Petersen E, Pawlowski ZS, Szczapa J. Neonatal screening for congenital toxoplasmosis in the Poznań region of Poland by analysis of Toxoplasma gondii-specific IgM antibodies eluted from filter paper blood spots. *Pediatr. Infect. Dis. J.,* 19(1):30-6, 2000.

Pelloux, H., Brun, E., Vernet, G., Marcillat S., Jolivet M. et al. Determination of anti-*Toxoplasma gondii* immunoglobulin G avidity: adaptation to the Vidas system (bioMérieux). *Diagn. Microbiol. Infect. Dis.,* 32: 69-73,1998.

Pelloux, H, Weiss, J, Simon, J, Muet, F, Fricker-Hidalgo, H, Goullier-Fleuret, A,Ambroise-Thomas, P . A new set of primers for the detection of Toxoplasma gondii in amniotic fluid using polymerase chain reaction. *FEMS Microbiol. Lett.,* 138:11-5, 1996.

Petersen E, Eaton RB. Control of congenital infection with *Toxoplasma gondii* by neonatal screening based on detection of specific immunoglobulin M antibodies eluted from phenylketonuria filter-paper blood-spot samples. *Acta. Paediatric.,* 88(432):36-9, 1999.

Pinon, JM and Gruson N. Interest of ELISA specific and compared immunological profiles in the early diagnosis of congenital toxoplasmosis. *Lyon. Med.,* 248:27-30, 1982.

Podzamczer D, et al. 2000. Thrice-weekly sulfadiazine-pyrimethamine for maintenance therapy of toxoplasmic encephalitis in HIV-infectedpatients. *Eur. J. Clin. Microbiol.. Infect. Dis.,* 19:89 –95, 2000.

Potasman I, Pick N, SrugoI. Screening for neonatal toxoplasmosis. *N. Engl. J. Med.,* 331 (21):1459, 1994..

Remington, JS, Araujo FG, and G. Desmonts. Recognition of diferents *Toxoplasma* antigens by IgM and IgG antibodies in mothers and their congenitally infected newborns. *J. Infect.Dis.,* 152:1020-1024, 1985.

Remington JS, Thulliez P, and. Montoya JG. Journal of Clinical Microbiology. Minireview. *Recent Developments for Diagnosis of toxoplasmosis,* Mar: 941-945, 2004.

Remington JS, McLeod R, Thulliez P, Desmonts G. Toxoplasmosis. In JS Remington, JO Klein, Infectious Diseases of the Fetus and Newborn Infant,5$^{th}$ ed. W.B. Saunders, Philadelphia, Pa. p. 947-1091, 2005.

Rey L. Toxoplasmose gondii e Toxoplasmose. In: Parasitologia. Rio de Janeiro: *Guanabara Koogan,* p: 321-334, 2001.

Robert-Gangneux F, Pharm FD, Peggy D, Chantal Y, Dorothé Q et al. Clinical relevance of placenta examination for the diagnosis of congenital toxoplasmosis. *Pediatr. Infect. Dis. J.,* 29:33–38, 2010.

Romand, S M, Wallon J, Franck, P, Thulliez, F, Peyron, and H. Dumon. Prenatal diagnosis using polymerase chain reaction on amniotic fluid for congenital toxoplasmosis. *Obstet. Gynecol.,* 97:296–300, 2001.

Roman S, Chosson M, Franck J, Wallon M, Kieffer F, Kaiser K, et al. Usefulness of quantitative polymerase chain reaction in amniotic fluid as early prognostic marker of fetal infection with Toxoplasma gondii. *American J. Obst. Gynecol.,* 190:797-802, 2004.

Schoen EJ, BlackS,Cohen D. Screening for neonatal toxoplasmosis. *N. Engl. J. Med.,* 331(21):1458,1994.

SU, C., Khan, A., Zhou, P, Majumdar, D, Ajzenberg, D, Dardé, ML, Zhu, XQ, Ajioka, JW, Rosenthal, BM, Dubey, J.P., Sibley, L.D. Globally diverse *Toxoplasma gondii* isolates comprise six major clades originating from a small number of distinct ancestral lineages. *Proc. Natl. Acad. Sci. U S A,* 109: 5844–5849, 2012.

Talabani H, Assef M, Year H, Delair E, Ancelle T, Thulliez P, Brézin AP, Dupouy-Camet J: Contributions of immunoblotting, real-time PCR, and the Goldmann-Witmer coefficient to diagnosis of atypical toxoplasmic retinochoroiditis. *J. Clin. Microbiol.,* 47:2131–2135, 2009.

Tissot Dupont, D, Fricker-Hidalgo H, Brenier-Pinchart MP, Bost-Bru C, Ambroise-Thomas P, and H. Pelloux. Usefulness of Western blot in serological follow-up of newborns suspected of congenital toxoplasmosis. *Eur. J. Clin. Microbiol., Infect. Dis.,* 22:122–125, 2003.

Van de Ven, E, MelchersW, Galama J, Camps W, and Meuwissen J. Identification of Toxoplasma gondii infections by BI gene amplification. *J. Clin. Microbiol.,* 19:2120–2124, 1991.

Verhofstede, C, Reniers, S, Colebunders, R, van Wanzeele, F, Plum, J Polymerase chain reaction in the diagnosis of toxoplasma encephalitis. *AIDS,* (7): 11, p. 1539-41, 1993.

Villard O, Filiisetti D, Roth-Deries F, Garweg J, Flament J et al. Comparison of enzyme-linked immunosorbent assay, immunoblotting, and PCR for diagnosis of toxoplasmic chorioretinitis. *J. Clin. Microbiol.,* 41: 3537–3541, 2003.

Wang, L, Chen, H, Liu, D, Huo, X, Gao, J, Song, X, Xu, X, Huang, K, Liu, W, Wang, Y, Lu, F, Lun, ZR, Luo, Q, Wang, X, Shen, J Genotypes and Mouse Virulence of Toxoplasma gondii Isolates from Animals and Humans in China. *PLoS One,* 8(1):e53483, 2013.

Weiss LM, Dubey JP. Toxoplasmosis: a history of clinical observations. *Int. J. Parasitol., 39*:895–901, 2009.

Wong, Sin-Yew, Remington, Jack S. Toxoplasmosis in pregnancy. Clin. Infect.Dis., 18:853-862,1994. World Health Organization. Toxoplasmosis: Report of a WHO Meeting of Investigators. Geneva. P.1-31 (Technical Report Series ,431), 1969.

World Health Organization. Toxoplasmosis: Report of a WHO Meeting of Investigators. Geneva. P.1-31 (Technical Report Series, 431), 1969.

In: *Toxoplasma Gondii*
Editors: M. Goodson and D. Winslow
ISBN: 978-1-62618-349-0


*Chapter II*

# Congenital Transmission of *Toxoplasma Gondii* in Ovine

***Thaís Rabelo dos Santos*[1], *Rodrigo Rabelo dos Santos*[1], *Alvimar José da Costa*[1] and *Katia Denise Saraiva Bresciani*[2]**
[1]FCAV, UNESP - Jaboticabal, Brazil
[2]Department of Production and Animal Health, FMVA, UNESP
Jaboticabal, SP, Brazil

## Abstract

*Toxoplasma gondii* is a member of the phylum Apicomplexa, order Coccidia, and is an obligate intracellular protozoan parasite that infects humans and several vertebrate hosts. In its life cycle, *T. gondii* has three infectious forms: the sporozoites (oocysts), tachyzoites (individually or in groups) and bradyzoites (tissue cysts). Toxoplasmosis is a zoonosis of major spread worldwide. Sheep toxoplasmosis was first described by Olafson and Monlux (1942) in the United States.These authors reported this disease in a female who presented neurological symptoms, hyperthermia, muscle rigidity, and other clinical signs. The diagnosis was made after necropsy and histopathological examination of the brain and spinal cord of the animal. In small ruminants, toxoplasmosis is relatively common, responsible for reproductive problems, generating great economic losses in sheep flocks, where the infection is the main cause of abortions, fetal malformations, premature births and neonatal deaths being the main route of infection the ingestion of sporulated oocysts of the parasite.

# Introduction

*Toxoplasma gondii* is a member of the phylum Apicomplexa, order Coccidia, and is an obligate intracellular protozoan parasite (Rorman et al., 2006) that infects humans and several vertebrate hosts (Dubey, 2004; Sharif et al., 2007).

The cats, particularly domestic cats, are the definitive hosts of the parasite (Hutchison, 1965; Faria et al., 2007) while warm blooded animals are intermediate hosts (Figure 1) (Dubey, 2004).

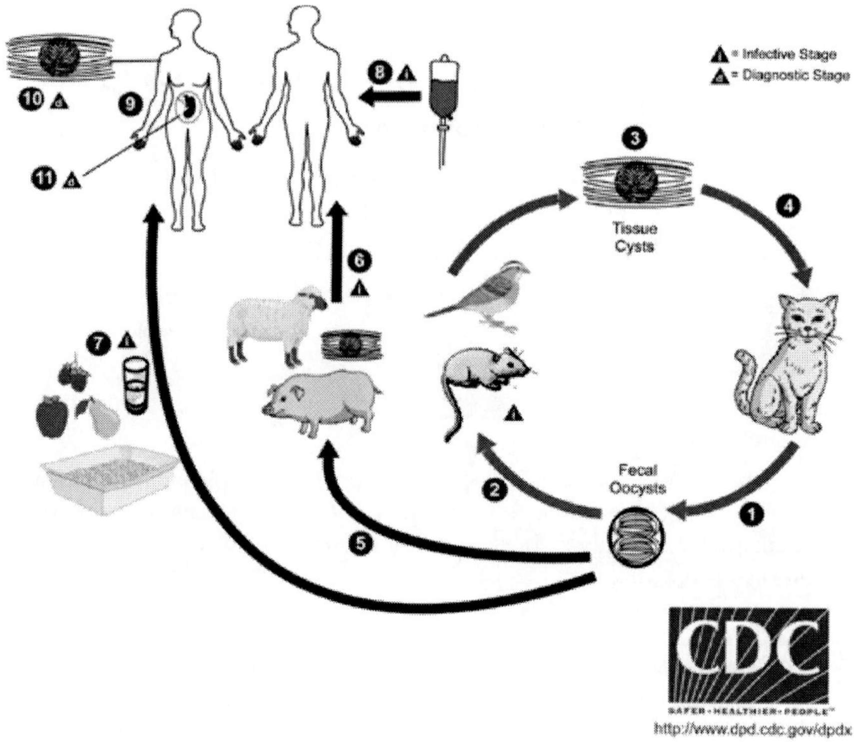


Figure 1. Cycle of *Toxoplasma gondii*, with its various forms of infection for humans (CDC, 2011).

In its life cycle, *T. gondii* has three infectious forms: the sporozoites (oocysts), tachyzoites (individually or in groups) and bradyzoites (tissue cysts) (Miller et al., 1972).

The cats may become infected by ingesting any of the three evolutionary forms, ie tachyzoites, cysts or oocysts. Particularly, the eating of tissue cysts, by carnivorism, is the most common pathway (Swango et al., 1989), which induces a lower prepatent period, and a greater production and excretion of oocysts in the feces compared to other forms of infection.

After ingestion of tissue cysts by cats, characterizing the enteroepitelial cycle (Wong and Remington, 1993), the cyst wall is digested by proteolytic enzymes in the stomach and small intestine and bradyzoites are released. Some penetrate the lamina propria of the intestine and multiply in tachyzoites, which in a few hours can be found in various tissues. Other bradyzoites penetrate the epithelial cells of the small intestine and initiate asexual development in many generations (schizonts types A-E) (Dubey and Frenkel, 1972).

Sexual reproduction occurs when organisms (merozoites) released from schizonts form male and female gametes. After fertilization of the female gamete, the formation of the oocyst wall starts, which upon reaching maturity, promotes disruption of intestinal epithelial cell then releasing the oocysts into the lumen (Dubey, 2004).

A single cat can release 20 million oocysts per day for approximately 20 grams of faeces (Fayer, 1981) and, after decomposition of the feces, soil contamination may be of 100,000 oocysts/gram (Frenkel et al., 1995), by a period from seven to 21 days. The oocysts are released in faeces in a non-sporulating form, therefore not infective (Frenkel et al., 1970; Fialho and Araujo, 2003). Sporulation occurs on the environment and ranges from one to five days, depending on the temperature and amount of oxygen. All cats are susceptible to infection by *T. gondii,* although younger animals, with up to one year of age, primoinfected, dispose a larger number of oocysts in the faeces after they have ingested tissue cysts containing the agent.Adult cats, primoinfecteds, also eliminate oocysts, but in a lower amount and for a shorter period (Lindsay et al., 1997).

Although most cats who had contact with *T. gondii* develop immunity and do not shed more oocysts, Dubey and Frenkel (1974) proved that seropositive cats, facing immunosuppression cases, may restart shedding oocysts. Later, Dubey (1995) reinfected seropositive cats, six years after they seroconvertion, and 55% of these eliminated oocysts in the feces again, but in much lower quantities to the primoinfection.

The fact that cats are the definitive hosts of *T. gondii,* was reinforced by Tenter et al. (2000) that reported epidemiological evidence in islets and atolls that were never inhabited by cats, were no infections by this pathogen existed.

The agent uses of surface structures and the apical complex to penetrate into the host cell. By joining the cell, by surface receptors (SAG-1), the protozoan directs its apical complex to create a intracellular junction, forming a vacuole (Huynh et al., 2003). This manner uses a minimal invasion of antigen exposure, which complicate the recognition by the immune system (Hiramoto et al., 2001). Thus, the parasite replicates and spreads rapidly to various tissues, such as the central nervous system (CNS), eye, placenta, skeletal and cardiac muscle, through the bloodstream (Montoya and Liensenfeld, 2004).

Some tachyzoites, after the host cell invasion, grow more slowly, forming bradyzoites, which will give origin to the cysts. The bradyzoites are morphologically similar to tachyzoites, however multiply slowly, express stage-specific molecules and are functionally different (Montoya and Liensenfeld, 2004), since the tissue cysts contain hundreds or thousands of bradyzoites (Dubey et al. 1998).

## Importance and Transmission to Humans

In animals destined for human consumption, 9.6% of the pigs (Suaréz-Aranda et al., 2000), 19.25% of cattle and 24.5% of sheep (Hashemi-Fesharki, 1996), were positive for *T. gondii,* which demonstrates the importance of meat consumption as a source of human contamination, if there is no adequate cooking of food (Dubey, 1996).In Europe, over 50% of sheep and pigs, researched by Acha and Szyfres (1986) were parasitized by *T. gondii.*

Toxoplasmosis is a zoonosis of major spread worldwide (Clementino et al., 2007), but the frequency of infection began to be characterized from 1948, with Sabin and Feldman, and then with other researchers, to develop the first diagnostic evidence of the disease .

Dubey (1986) establishes the transmission of this disease, primarily in three ways: fecal-oral, congenital and carnivorism .

Regarding the fecal-oral route, the cats excrete, for a period of four to 14 days, with the faeces, non sporulated *T. gondii* oocysts and, therefore, not infectious.The sporulation process may take from one to five days, depending on the ambient conditions, after defecation, when these oocysts become infective to other species (Dubey, 1994).

In carnivorism, infection occurs by ingestion of cysts containing bradyzoites and/or tachyzoites present in the meat of an intermediate host (Dubey, 1996). In relation to the resistance of the cysts in meat, it is believed that they do not resist to the processes of salting and heating (above 55ºC for more than two minutes), used in the manufacture of processed meat (Dubey, 1994). However, other studies show that the process of seasoning, sausages and salami, mostly, and cooled meat, may represent an important route of transmission of toxoplasmosis and may also be responsible for the high prevalence of toxoplasmosis eye injuries in southern Brazil (Glasner et al., 1992, Arias et al., 1996).

Infection by meat can happen not only by consumption, but also by manipulating the same raw food during preparation, contact with food preparation surfaces, knives and other utensils (Smith et al., 1992). Recent studies by Liu et al., (2009) showed that the act of pregnant women experience the seasonings for raw meat, while preparing the food, is one of the major risk factors for these women to acquire toxoplasmosis during pregnancy.

There is also the possibility of the disease spread in zoos with the equipment of daily use, concurrently, in the grounds of felids, such as boots, gloves, shovels, brooms and hoses (Dubey, 1986).

Another acknowledged route of transmission is the contact with dogs, due the behavior of some of them to roll over and eat cat feces, perhaps attracted by the smell. This behavior was called xenosmofilia and would enable the transmission by contact with sporulated oocysts present in the coat of dogs living in highly contaminated areas (Frenkel, and Parker, 1996).

Direct contact with felines presents epidemiological little consequence, however the presence of infected animals may indicate a contaminated environment, bringing risk to the human population and other animals (Santos et al., 2009).

Less common, but possible, infection can occur through ingestion of unpasteurized milk with tachyzoites, as described in goats by Chiari et al. (1984), laboratory accidents (Wong and Remington, 1993), blood transfusion (Figueroa-Damian, 1998) and organ transplantation (Munir et al., 2000). In women, although there is no proven transmission through breastfeeding, Bonametti et al. (1997) described a positive case of toxoplasmosis in a child whose diet consisted exclusively of breast milk, and the mother was part of a group of 17 individuals who had acquired toxoplasmosis by eating raw lamb. In this case, a serological survey on the property of origin of sheep revealed

43% of toxoplasmosis positive animal. The risk of acquiring this disease in postnatal life was also demonstrated by Dubey et al. (1998), Giraldi et al. (2001), Silva et al. (2001), Kravetz Federman, 2005; Bachmeyer et al., (2006) and Hung et al., (2007).

In humans, infection is very common, but the clinical disease is less frequent. It is estimated that one third or more, of the population have antibodies against the parasite. Only in the U.S. is likely to occur 1,437,500 new cases of infection with *T. gondii* (Todd, 1989).Still in the U.S., according to Saad et al. (1996), approximately 300 people die annually due to toxoplasmosis, both in its congenital form (neonates) and in individuals with severe iatrogenic immunosuppression (transplanted, treatment for cancer, patients undergoing renal dialysis) or suffering from immunosuppressive diseases such as autoimmune diseases or acquired immunodeficiency syndrome - AIDS (Smith, 1993).

In patients with HIV, toxoplasmosis manifests itself with a high frequency and is considered an opportunistic disease (Lutf and Remington, 1988) which causes the reactivation of cysts, especially the ones in the brain, producing severe encephalitis (Halonen et al., 2001 and Khetsuriani et al., 2002; Mamidi et al., 2002; Collazos, 2003; Yadav et al., 2004; Pradhan et al., 2007). Potasman et al. (1987) observed a frequency of encephalitis *T. gondii* between six and 10% of AIDS patients in the U.S. In Brazil, in patients above 12 years of age, toxoplasmosis is the third most prevalent associated disease with 11,809 (3.72%) of the 314,582 cases reported between 1980 and 1996 (Brazil, 1996).

# Toxoplasmosis in Sheep

Sheep toxoplasmosis was first described by Olafson and Monlux (1942) in the United States.These authors reported this disease in a female who presented neurological symptoms, hyperthermia, muscle rigidity, and other clinical signs. The diagnosis was made after necropsy and histopathological examination of the brain and spinal cord of the animal .In small ruminants, toxoplasmosis is relatively common (Mainar et al., 1996), being responsible for reproductive problems, generating great economic losses in sheep flocks (WYSS et al., 2000; Pereira-Bueno et al., 2004), where the infection is the main cause of abortions, fetal malformations, premature births and neonatal deaths (Malık et al., 1990), being the main route of infection the ingestion of sporulated oocysts of the parasite (Dubey and Beverley, 1988). Infection acquired in the beginning of pregnancy (prior to 50 days), before the fetus develop the capacity of producing antibodies normally cause embryonic death and resorption (Hartley, 1961). If the sheep become infected with *T. gondii* mid-pregnancy (70-90 days) there is a considerable probability of miscarriage or stillbirth (Beverley and Watson, 1971; Watson and Beverley, 1971, Miller et al. 1 982), while in late pregnancy (> 110 days) will litter normally sheep, lambs though their may be congenitally infected (Watson and Beverley, 1971; Miller et al. 1982).However, few studies have described the occurrence of healthy newborn lambs, but infected with *T. gondii* in populations of sheep (Duncanson et al. 2001). As a result, New Zealand and some European countries, has been using a licensed vaccine (attenuated strain "S48") of tachyzoites, with the goal of minimizing economic losses in these production systems (Montoya and Liesenfeld, 2004).

In Brazil, infection rates (Table 1) pointed to sheep flocks vary or oscillate from 8% (Nishikawaet al. 1,984) to 55% (Langoni et al. 1999).This fluctuation in occurrence is mainly due to the type of serologic test used, the region and the age of the animals studied (Dubey, 1990). The high prevalence of toxoplasmosis in sheep may be linked to lower resistance of this specie to the parasite and the conditions of sheep industry exploitation that provide greater risk of exposure and contact of these animals with oocysts eliminated by cats (Dubey and Hamir, 2002). Sposito-Filha et al. (1992) isolated 20 strains of *Toxoplasma* from the diaphragm muscle of 136 sheep.Da Silva and Langoni (2001) identified this protozoan in the brain and diaphragm of 34 of 40 sheep serologically reactive to indirect immunofluorescence assay. These works of isolating viable parasites in edible tissues and goat milk (Chiari and Neves, 1984) reveal the importance of these species as a source of infection for humans.

*T. gondii* infection, in humans, produces strong immunity against reinfection, limiting future congenital transmission during the next pregnancy, and any subsequently generated child will not be infected (Frenkel, 1990). Experimental vaccination using attenuated strain "S48" of *T. gondii* were successful in limiting the severity of congenital diseases, when sheep were challenged with oocysts (Buxton and Innes, 1995).However, despite the success of vaccination in limiting abortion and congenital disease, this does not necessarily blocks the transmission of the disease (Buxton and Innes, 1995). It also revealed that an oral dose of infective oocysts inoculated before pregnancy was sufficient to produce an immune response and therefore counteract the parasitic challenge (McColgan et al, 1988).

**Table 1. Occurrence (%) of antibodies against *Toxoplasma gondii* in sheep from different Brazilian regions, used techniques and respective authors in different studies**

| Region | State | Author | Number of examined animals | Techniques | Occurrence(%) |
|---|---|---|---|---|---|
| North | Rondônia | Cavalcante et al. (2004) | 141 | IFAT | 46.80 |
| Northeast | Bahia | Gondim et al. (1999) | 240 | IFAT | 18.75 |
| | Pernambuco | Silva et al. (2003) | 173 | IFAT | 35.30 |
| South | Rio Grande do Sul | Amaral et al. (1978) | 100 | SF | 23.00 |
| | | Larsson et al. (1980) | 100 | SF | 39.00 |
| | | Costa and Silva (1981) | 310 | IFAT | 12.00 |
| | | Nishikawa et al. (1984) | 655 | LAT | 8.00 |
| | | Silva And De La Rue (2006) | 87 | IFAT | 44.80 |
| | Paraná | Romanelli et al. (2007) | 305 | IFA | 51.47 |
| Southeast | São Paulo | Langoni et al. (1999) | 352 | IFA | 55.00 |
| | | Silva et al. (2002) | 100 | IFA | 23.00 |
| | | Meirelles (2003) | 200 | IFA | 31.00 |
| | | Figioulo et al. (2004) | 597 | IFA | 34.70 |
| | | Lopes et al. (2010) | 488 | IFA | 52.05 |
| Central West | Federal District | Ueno et al. (2009) | 1028 | IFA | 38.20 |

IFAT: Indirect Immunofluorescence Antibody Test, SF: Sabin and Feldman; LAT: Latex Agglutination Test.

The hypothesis that the primary infection protects against reinfection is the basis for many sheep farmers do not discard sheep with abortion history. However, recent studies suggested that sheep persistently infected with *T. gondii* can pass the infection congenitally more frequently than expected (Buxton et al., 2007).

The work of experimental infection (controlled and in conditions that more closely resemble the natural ones) are best suited to provide clinical, laboratorial and epidemiological data about the disease, especially in relation to prophylaxis.

Researchers from England (Duncanson et al., 2001; Morley et al. 2005; Williams et al. 2005; Morley et al., 2007) proposed that transplacental transmission repeated by *T. gondii* in sheep may be more common than previously believed. However, all the presented evidences were based on the detection of *T. gondii* DNA by PCR (Dubey, 2009).

Congenital transmission at high rates of occurrence (61%) was found in commercial herds (placenta and fetal tissues) using the Polymerase Chain Reaction for (PCR).The *T. gondii* was isolated in 94% of pregnancies that resulted in deaths and 42% of that came to an end (Duncanson et al., 2001).

Williams et al. (2005) detected high levels of congenital transmission in Charollais ewes (50.5%) and in commercial herds, crossbred (69%) by PCR. However it was not possible to prove if the congenital transmission ocurred in the first infection or with the chronic reactivation of the infection, during pregnancy, but the authors suggest that due to the high levels of transmission, there must have been the reactivation of toxoplasmosis. Buxton et al. (2006) and Silva and Rue (2006) reported the possibility of transmission of toxoplasmosis in sheep persistently positive, although little, shall not be discarded. Freyre et al. (2009) evaluating an experimental model of congenital transmission in hamster found that this route of transmission occurred in 13 out of 17 hamsters infected with bradyzoites or *T. gondii* oocysts.

As Morley et al. (2005), one hypothesis is that vertical transmission of *T. gondii* occurs from generation to generation after primary infection, resulting in the maintenance of high levels of infection in some families, while others do not.This pattern is mimicked by the occurrence of abortion in the same families due to the strong correlation between infection and abortion. The belief that the *T. gondii* can be transmitted to the offspring of infected family lines over successive generations is controversial, however should not be discarded.

According to Morley et al. (2007), there is a high risk (55%) of production of a stillbirth after a previous abortion, most likely associated with toxoplasmosis. It is also reported that a third part of ovine abortions diagnosed is related to *T. gondii* infection. The paper describes the need for further studies on acquired immunity in primary infection with *T. gondii* lifelong, besides its transmission to offspring.

Rodger et al. (2006) have failed to detect congenital transmission in persistently infected sheep.Thirty-one sheep naturally seropositive for *T. gondii* and 15 seronegative ewes were mated and controlled during pregnancy.It was found no evidence of toxoplasmosis in histopathology or PCR of placenta or lambs tissues. Low titers to the parasite were found in three lambs, but it was not possible to establish whether these antibodies represent evidence of fetal infection.

These studies demonstrate congenital transmission, however there wasn't a detailed study of primary infection, reactivation of chronic infection or reinfection during pregnancy.Moreover, all these studies are based on DNA detection by PCR of placenta and fetal tissues.

Recently, Santos (2012) in a study of experimental infection and reinfection (and controlled conditions that more closely resemble the natural) confirmed the congenital transmission in sheep persistently infected and re-infected sheep at different gestational stages, *Toxoplasma gondii.* The acquired immunity in primary infection did not protect sheep against future reinfection by heterologous strains of *Toxoplasma gondii.* In this study, the tissue parasitism by *T. gondii* could be diagnosed in females (infected and reinfected) and their lambs, through bioassay in mice, immunohistochemistry and PCR . Therefore, the results obtained in this recent study confirmed the findings reported by Duncanson et al. (2001), Morley et al. (2005), Williams et al. (2005), Morley et al. (2007).

# References

Acha, PN. and Szyfres, B. Zoonosis y enfermidades transmisibles comunes al hombre y animales. 2.ed. Washington: *Organization Panamericana de la Salud*, 1986, (Publicación Científica 503).

Amaral, V.; Santos, SM.; Rebouças, MM. Sobre a prevalência de anticorpos anti-*Toxoplasma* em soros de caprinos e ovinos procedentes, respectivamente, dos Estados da Bahia e Rio Grande do Sul, Brasil. *O Biológico,* 1978, v.12, p.331-340.

Arias, MA.; Chinchilla, M.; Reyes, L.; Linder, E. Seroepidemiology of Toxoplasmosis in humans posible transmission routes in Costa Rica. *Revista de Biologia Tropical*, 1996, v. 44, p. 377-381.

Bachmeyer, C.; Mouchnino, G.; Thulliez, P.; Blum, L. Congenital toxoplasmosis from an HIV-infected woman as a result of reactivation. *The Journal of Infection*, 2006, v. 52, p.e55–e57.

Beverley JKA., Watson WA., Spence, JB. The pathology of the foetus in ovine abortion due to toxoplasmosis. *Veterinary Record*, 1971, v.88, p.174-178.

Bonametti, AM.; Passos, JN.; Silva, EMK. et al. Surto de toxoplasmose aguda transmitida através da ingestão de carne crua de gado ovino. *Revista da Sociedade Brasileira de Medicina Tropical*, 1997, v.30, p.21-25.

Brazil. Ministério da Saúde. Fundação Nacional de Saúde. Boletim Eletrônico Epidemiológico, v.6, n.3, 2006.

Buxton, D.; Innes, EA. A commercial vaccine for ovine toxoplasmosis. *Parasitology,* 1995, 110 (Suppl.), p.S11-S16.

Buxton, D.; Rodge, SM.; Maley, SW.; Wright, SE. Toxoplasmosis: The possibility of vertical trnasmission. *Small Ruminant Research, 2006,* v.62, p.43-46.

Buxton, D.; S. Maley, W.; Wright, SE.; Rodger, S.; Bartley, P.; INNES, EA. *Toxoplasma gondii* and ovine toxoplasmosis: New aspects of an old story. *Veterinary Parasitology*, 2007, v.149, p.25-28.

Cavalcante, GT., Aguiar, DM., Chiebao, DP., Meireles, LR., Andrade, HF., Camargo, LMA., Labruna, MB., Ruiz, VLA., Gennari, SM. Ocorrência de anticorpos anti-*Toxoplasma gondii* em humanos e animais domésticos da zona Monte Negro, Rondônia. Thirteenth Congresso Brasileiro de Parasitologia Veterinária, Ouro Preto, Minas Gerais, vol. 13 (Suplemento), Brasil, 20-24 de setembro de 2006, *Revista Brasileira Parasitologia Veterinária*, 2004, 217.

CDC, 2011. Disponível em: http://www.cdc.gov/parasites/toxoplasmosis/ biology.html Acesso em 14 de dezembro de 2011.

Chiari, CA.; Neves, DP.; PEREIRA NEVES, D. Toxoplasmose humana adquirida através da ingestão de leite de cabra. *Memórias do Instituto Oswaldo Cruz*, 1984, v.79, n.3, p.337-340.

Clementino, MM.; Souza, MF.; Andrade Neto, VF. Soroprevalence and *Toxoplasma gondii*-IgG avidity in sheep from Lajes, Brasil. *Veterinary Parasitology,* 2007, v.146, p.199-203.

Collazos, J. Opportunistic infectious of the CNS in patients with AIDS: diagnosis and management. *CNS Drugs,* 2003, v.17, n. 12, p.869-887.

Da Silva, AV.; Langoni, H. The detection of *Toxoplasma gondii* by comparing cytology, histopathology, bioassay in mice, and the polymerase chain reaction (PCR). *Veterinary Parasitology,* 2001, v.97, n.3, p.191-198.

Dubey, JP.,. Toxoplasmosis - a waterborne zoonosis. *Vet. Parasitol.* 2004, v. 126, p.57–72.

Dubey, JP. Toxoplasmosis. *Journal of the American Veterinary Medical Association,* 1986, v.189, n.2, p.166-170.

Dubey, JP. Toxoplasmosis. *Journal of the American Veterinary Medical Association*, 1994, v.205, n.11, p.1593-1598.

Dubey, J.P. Refinement of pepsin digestion method for isolation of *Toxoplasma gondii* from infected tissues. *Veterinary Parasitology*, 1998, v.74, p.75-7.

Dubey, JP. Toxoplasmosis – a waterborne zoonosis. *Veterinary Parasitology,* 2004, v.126, p.57-72.

Dubey, JP. Toxoplasmosis. *Journal of the American Veterinary Medical Association*, 1986, v.189, n.2, p.166-170.

Dubey, JP. Duration of immunity to shedding of *Toxoplasma gondii* oocysts by cats. *Journal of Parasitology*, 1995, v. 81, n. 3, p. 410-415, 1995.

Dubey, JP. e Beverley, JKA.Toxoplasmosis of animals and man. *Boca Raton: Academic*,1988, 315p.

Dubey, JP. Pathogenicity and infectivity of *Toxoplasma gondii* oocysts for rats. *Journal of Parasitology*, 1996, v.82, p.951-956.

Dubey, JP. Toxoplasmosis in sheep — The last 20 years. *Veterinary Parasitology*, 2009, v.163, p.1-14.

Dubey, JP. Toxoplasmosis. *Journal of the American Veterinary Medical Association,* 1990, v.205, n.11, p.1593-1598.

Dubey, JP. Toxoplasmosis. *Journal of the American Veterinary Medical Association*, 1994, v.205, n.11, p.1593-1598.

Dubey, JP., Frenkel, JK. Cyst-induced toxoplasmosis in cats. *Journal of Protozoology*, 1972, v.19, p.155-177.

Dubey, JP., Hamir, AN. Experimental toxoplasmosis in budgerigars (*Melopsittacus undulatus*). *Journal of Parasitology, 2002,* v.88, p.514-9.

Dubey, JP.; Frenkel, JK. Immunity to feline toxoplasmosis: modification by administration of corticosteroids. *Veterinary Pathology*, 1974, v. 11, p. 350-379.

Duncanson, P.; Terry, RS.; Smith, JE; Hide, G. High levels of congenital transmission of Toxoplama gondii in a commercial sheep flock, *International Journal for Parasitology*, 2001, v.31, n.14, p.1699-1703.

Duncanson, P.; Terry, RS.; Smith, JE; Hide, G. High levels of congenital transmission of Toxoplama gondii in a commercial sheep flock. *International Journal for Parasitology,* 2001, v.31, n.14, p.1699-1703.

Faria, EB.; Gennari, SM.; Pena, HFJ.; Athayde, ACR.; Silva, Mlcr.; Azevedo, SS. Prevalence of anti-*Toxoplasma gondii* and anti-*Neospora caninum* antibodies in goats slaughtered in the public slaughterhouse of Patos city, Paraíba State, Northeast region of Brazil. *Veterinary Parasitology,* 2007, v.149, p.126–129.

Fayer, R. Toxoplasmosis update and public health implications. *The Canadian Veterinary Journal*, 1981, v. 22, p.344-352.

Fialho, CG. And Araujo, FAP. Detecção de anticorpos para *Toxoplasma gondii* em soro de suínos criados e abatidos em frigoríficos da região da grande Porto Alegre – RS, Brasil. *Ciência Rural*, 2003, v.33, n.5, p.893-897.

Figueroa Damian, R. Risk of transmission of infectious diseases by transfusion. *Ginecol Obstet Mex.*, 1998, v. 66, p. 277-283.

Frenkel, JK. Toxoplasmosis in human beings. *Journal of The American Veterinary Medical Association*, 1990, Schaumburg, v. 196, n.2, p. 240-248.

Frenkel, JK.; Dubey, JP.; Miller, NL. *Toxoplasma gondii* in cats: fecal stages identified as coccidian oocysts. *Sciences*, 1970, v. 167, p. 893-896.

Frenkel, JK.; Hassanein, KM.; Hassanein, RS.; Brown, E.; Thulliez, P.; Quiteronunez, R. Transmission of *Toxoplasma gondii* in Panama-City. *The American Journal of Tropical Medicine and Hygiene,* 1995, v.53, p. 458-468.

Frenkel, JK.; Parker, BB. An apparent role of dogs in the transmission of *Toxoplasma gondii.* The probable importance of Xenosmophilia. *Annals of the New York Academy of Sciences*, 1996, v. 791, p. 402-407.

Freyre, A.; Fialho, CG.; Bigatti, LE.; Araujo, FAP.; Falcón, JD.; Mendez, J.; González, M. *Toxoplasma gondii*: Congenital transmission in a hamster model. *Experimental Parasitology,* 2009, v. 122, p. 140-144.

Giraldi, N., Vidotto, O., Navarro, IT., Garcia, JL., Ogawa, L., Kobyka, E. Toxoplasma antibody and stool parasites in public school children, Rolândia, Paraná, Brazil. *Revista da Sociedade Brasileira de Medicina Tropical*, 2001, v.35 n.3.

Glasner, PD.; Silveira, C.; Kruszon-Moran, D.; Mertins, MC.; Burner, M.; Silveira, S.; Camargo, ME.; Nussenblet, RB.; Kaslow, RA.; Belfort, R. An Unsually High prevalence of Ocular toxoplasmosis in Southern Brazil. *American Journal of Ophthalmology*, 1992, v. 114, p. 136-144.

Gondim LFP., Barbosa JR, HV., Ribeiro Filho, CHA., Saeki, H. Serological survey of antibodies to *Toxoplasma gondii* in goats, sheep, cattle and water buffaloes in Bahia State, Brazil. *Veterinary Parasitology*, 1999, v.82, p.273-276.

Halonen, SK., Taylor, GA., Weiss, LM. Gamma interferon- induced anhibition of *Toxoplasma gondii* in astrcytes is mediated by IGIP. *Infection and Immunity*, 2001, v. 69, p. 5573-6.

Hartley, WJ.; Moyle, GG. Further observations on the epidemiology of ovine *Toxoplasma* infection. *Australian Journal of Experimental Biology and Medical Science*, 1974, v.52, p.647-53.

Hashemi-Fesharki, R. Seroprevalence of *Toxoplasma gondii* in cattle, sheep and goats in Iran. *Veterinary Parasitology,* 1996, v.61, n.1-2, p.1-3.

Hiramoto, RM. ; Mayrbaurl-Borges, M.; Galisteo JR, AJ.; Meireles, L. R.; Macre, M. S; Andrade JR, HF. Infectivity of cysts of the ME-49 *Toxoplasma gondii* strain in bovine milk and homemade cheese. *Rev Saúde Pública*, 2001, v.35, n.2, p.113-118.

Hung, CC.; Fan, CK.; Su, KE.; Sung, FC.; Chiou, HY.; Gil. V.; Ferreira, MDAC.; Carvalho, JM.; Cruz, C.; Lin, YK.; Tseng, LF.; Sao, KY.; Chang, WC.; Lan, HS.; Chou, SH. Serological screening and toxoplasmosis exposure factors among pregnant women in the Democratic Republic of Sao Tome and Principe. *Transactions of the Royal Society of Tropical Medicine and Hygiene*, 2007, v. 101, n.2, p. 134-139.

Hutchison, WM. Experimental transmission of *Toxoplasma gondii. Nature*, 1965, v.206, p.961-962.

Huynh, MH., Rabenau, KE., Harper, JM., Beatty, WL., Sibley, LD., Carruthers, VB. Rapid invasion of host cells by Toxoplasma requires secretion of the MIC2-M2AP adhesive protein complex. *EMBO Journal*, 2003, v.22, p.2082-2090.

Khetsuriani, N., Holaman, RC., Anderson, LJ. Burden of encephalitis-associated hospitalizacions in the United States, 1988-1997. Clinical Infectious Diseases, 2002, v.35, n.2, p. 175-82.

Kravetz, JD.; Federman, DG. Toxoplasmosis in pregnancy. *The American Journal of Medicine,* 2005, v. 118, n. 3, p. 212-216.

Langoni, H. et al. Inquérito soroepidemiológico para a toxoplasmose em ovinos no Estado de São Paulo, Brasil. *O Biológico*, 1999, São Paulo, v.61, n.1, p.35-39.

Larsson, CE.; Jamra, LMF.; Guimarães, EC. Prevalência da toxoplasmose ovina determinada pela reação de Sabin-Feldman em animais de Uruguaiana, RS, Brasil. *Revista de Saúde Pública,* 1980, v.14, p.582-588.

Lindsay, DS.; Dubey, JP., Butler, JM., Blagburn, BL. Mechanical transmission of *Toxoplasma gondii* oocysts by dogs. *Veterinary Parasitology*, 1997, v.73, n.1, p.27-33.

Liu, Q.; Wei F.; Gao S.; Jiang L.; Lian H.; Yuan B.; Yuan Z.; Xia Z.; Liu B.; Xu X.; Zhu, XQ. *Toxoplasma gondii* infection in pregnant women in China. *Transactions of the Royal Society of Tropical Medicine and Hygiene,* 2009, v. 103, p. 162-166.

Lopes, WDZ., Santos, TR., Santos, RS., Rossanese, WM., Souza, FA.; Rodrigues, JDF.; Mendonça, RP.; Soares, VE.; Costa, AJ. Seroprevalence of and risk factors for *Toxoplasma gondii* in sheep raised in the Jaboticabal microregion, São Paulo State, Brazil. *Research Veterinary Science*, 2010, v.88, p.104-106.

Luft, RZ.: Remington, JS. Toxoplasmic encephalitis . *The Journal of Infectious Diseases*., 1988, v.157, p.01-06.

Mainar, RC., Cruz, C. De La, Asensio, A., Domínguez, L., Vázquez-Boland, JA. Prevalence of agglutinating antibodies to *Toxoplasma gondii* in small ruminants of the Madrid region, Spain, and identification of factors influencing seropositivity by multivariate analysis. *Veterinary Research Communications*, 1996, v.20, p.153–159.

Malik, MA., Deesen DW.; Cruz, A. Toxoplasmosis in sheep in northeastern United States. *Journal of the American Veterinary Medical Association*, 1990, v.196, p.263-265.

Mamidi, A., De Simone, JA., Pomerantz, RJ. Central nervous system infectious in individuais whith HIV-1 infection. *Journal For Neurovirology*, 2002, v.8, n.3, p.158-67.

Mccolgan, C., Buxton, D., Blewett, DA. Titration of *Toxoplasma gondii* oocysts in non-pregnant sheep and the effects of subsequent challenge during pregnancy. *Veterinary Record*, 1988, v.123, p.467–470.

Meireles, LR.; Galisteo JR, AJ.; Andrade JR., HF. Serological survey of antibodies to *Toxoplasma gondii* in food animals from São Paulo state, Brazil. *Brazilian Journal of Veterinary Research and Animal Science*, 2003, v.40, p.267-271.

Miller, JR., Blewett, DA., Buxton, D., Clinical and serological response of pregnant gimmers to experimental induced toxoplasmosis. *Veterinary Record*, 1982, v.111,p.124-126.

Miller, NL.; Frenkel, JK.; Dubey, JP. Oral infections with Toxoplasma cysts and oocysts in felines, other mammals and birds. *Journal of Parasitology*, 1972, Lawrence, v.58, n.5, p.928-937.

Montoya, JG.; Liesenfeld, O. Toxoplasmosis. *Lancet, London*, 2004, v.363, n.9425, p.1965-1976,.

Morley, EK., Williams, RH., Hughes, JM., Thomasson, D., Terry, RS., Duncanson, P., Smith, JE., Hide, G. Evidence that primary infection of Charollais sheep with *Toxoplasma gondii* may not prevent foetal infection and abortion in subsequent lambings. *Parasitology,* 2008v.135, p.169-173.

Morley, EK.; Williams, RH.; Hughes, JM.; Terry, RS.; Duncanson, P.; Hide, G. Significant familial differences in the frequency of abortion and *Toxoplasma gondii* infection within a flock of Charollais sheep. *Parasitololgy,* 2005, v. 131, p. 181-185.

Munir, A.; Zaman, M.; Eltorky, M. *Toxoplasma gondii* pneumonia in a pancreas transplant patient. *The Southern Medical Journal*, 2000, v.93, n.06, p.614-617.

Nishikawa, H.; Arnoni, JV.; Rassier. DSS.; Silva, SS. Prevalence of antibodies to *Toxoplasma gondii* in domestic animals in Rio Grande do Sul State, Brazil. *Resumos I Encontro de Pesquisas Veterinárias*, 1984, Universidade Estadual de Londrina, Paraná, 62p.

Olafson, P. and Monlux, WS. *Toxoplasma* infection in animals. *Cornel. Vet.* 32, 316-6, 1942.

Pereira-Bueno, J.; Quintanilla-Gonzalo, A.; Pérez-Perez, V.; Álvarez-Garcia, G.; Collantes-Fernández, E.; Ortega-Mora, LM. Evaluation of ovine abortion associated with *Toxoplasma gondii* in Spain by different diagnostic techniques. *Veterinary Parasitology,* 2004, v.121, p.33-43.

Potasman, I.; Araujo, FG.; Thulliez, P.; Desmonts, G.; Remington, JS. *Toxoplasma gondii* antigens recognized by sequential sample of serum obtained from congenitally infants. *Journal of Clinical Microbiology*, 1987, v. 25, n. 10, p. 1926-1931.

Pradhan, S.; Yadav, R.; MISHRA, VN. *Toxoplasma* meningoencephalitis in HIV-seronegative patients: clinical patterns, imaging features and treatment outcome. *Transactions of the Royal Society of Tropical Medicine and Hygiene*, 2007, v.101, p.25-33.

Rodger, SM., Maley, SW., Wright, SE., Mackellar, A., Wesley, F., Sales, J., Buxton, D. Role of endogenous transplacental transmission in toxopasmosis in sheep. *Veterinary Record*, 2006, v.159, p.768–772.

Romanelli, PR., Freire, RL., Vidotto, O., Marana, ERM., Ogawa, L., De Paula, VSO., Garcia, JL., Navarro, IT. Prevalence of Neospora caninum and *Toxoplasma gondii* in sheep and dogs from Guarapuava farms Prevalence of Neospora caninum and *Toxoplasma gondii* in sheep and dogs from

Guarapuava farms, Parana' State, Brazil. *Research Veterinary Science*, 2007, v.82, p.202–207.

Rorman, E., Zamir, CS., Rilkis, I., Ben-David, H., 2006. Congenital toxoplasmosis—prenatal aspects of Toxoplasma gondii infection. *Reprod.Toxicol.* 21, 458–472.

Saad, R.; Vincent, JF.; Cimon, B.; De Gentile, L.; Francois, S.; Bouachour, G.; Ifrah, N. Pulmonary toxoplasmosis after allogeneic bone marrow transplantation: case report and review. *Bone Marrow Transplant*, 1996, v.18 p. 211-212.

Sabin, AB. and Feldman, HA. Dyes as microchemical indicators of a new immunity phenomenon affecting a protozoon parasite (*T. gondii*). *Science*, 1948, Washington,v.108, p.660-663.

Santos, TR. 2012. *Toxoplasmose congênita em ovelhas reinfectadas experimentalmente*. 2012. 120F. Tese (Doutorado em Medicina Veterinária – área de concentração: Patologia Animal) – Faculdade de Medicina Veterinária de Jaboticabal, UNESP, Jaboticabal, SP, 2012.

Santos TR., Costa AJ., Toniollo GH., Luvizotto MCR., Benetti AH., Santos RR., Matta DH., Lopes WDZ., Oliveira JA. and Oliveira GP. Prevalence of anti-*Toxoplasma gondii* antibodies in dairy cattle, dogs, and humans from the Jauru micro-region, Mato Grosso state, Brazil. *Veterinary Parasitology*, 2009, v.161, p.324-326.

Sharif, M., Gholami, S.H., Ziaei, H., Daryani, A., Laktarashi, B., Ziapour, S.P., Rafiei, A., Vahedi, M., Seroprevalence of Toxoplasma gondii in cattle, sheep and goats slaughtered for food in azandaran province, Iran, during 2005. *Vet. J.* 2007, v.174, p. 422–424.

Silva, AV.; Cutolo, AA.; Langoni, H. Comparação da reação de imunofluorescência indireta e do método de aglutinação direta na detecção de anticorpos anti-Toxoplasma em soros de ovinos, caprinos, caninos, e felinos. *Arquivos do Instituto Biológico*, 2002, v. 69, n. 1, p. 7-11.

Silva, KLM.; Rue, ML. Possibilidade da transmissão congênita de *Toxoplasma gondii* em ovinos através de seguimento sorológico no município de Rosário do Sul, RS, Brasil. *Ciência Rural*, 2006, v. 36, p. 892-897.

Silva, KLV e De La Rue, ML. Possibilidade de transmissão congênita de *Toxoplasma gondii* em ovinos através de seguimento sorológico no município de Rosário do Sul, RS, Brasil. (Possibility of *Toxoplasma gondii* congenital transmission in sheep by serological follow-up in Rosário do Sul County, RS, Brazil). *Ciência Rural*, 2006, v.36, p.892–897.

Silva, LA., Cunha, ELP., Meireles, LR., Gottschalk, Mota, RA., Langoni, H. Toxoplasmose em ovinos e caprinos estudo soroepidemiologico em duas regiões do Estado de Pernambuco, Brasil. *Ciência Rural*, 2003, v.33 p. 115-119.

Silva, LA., Vieira, RS., Serafini, LN., Junior, CGC., Figueiredo, JFC. Toxoplasmose do sistema nervoso central em pacientes sem evidência de imunosupressão: relato de caso. *Revista da Sociedade Brasileira de Medicina Tropical*, 2001, v.34, n.5.

Smith, BP. Tratado de medicina interna de grandes animais. *Editora Manole*, 1993, São Paulo. V.1,2, 1738p.

Smith, KE.; Zimmerman, JJ.; Patton, S.; Beran, GW.; Hill, HL. The epidemiology of toxoplasmosis on lowa swine farms with an emphasis on the roles of free-living mammals. *Veterinary Parasitology,* 1992, v. 42, p.199-211.

Spósito Filha, E.; Amaral, V.; Macruz, R.; Rebouças, MM.; Santos, SM.; Drumond, LS. *Toxoplasma gondii* em ovinos: Isolamento do parasita a partir de diafragmas de animais procedentes do Estado do Rio Grande do Sul e abatidos em matadouros de São Paulo, para consumo humano. *Revista Brasileira de Parasitologia Veterinária*, 1992, v.1, p.117-119.

Suaréz-Aranda, F.; Galisteo JR, AJ.; Hiramoto, RM.; Cardoso, RPA; Meireles, LR.; Miguel, O.; Andrade JR., HF. The prevalence and avidity of Toxoplasmagondii IgG antibodies in pigs from Brazil and Peru. *Veterinary Parasitology*, 2000, v. 91, n. 1-2, p. 23-32.

Swango, LJ.; Bankemper, KW.; Kong, LI. Bacterial, rickettsial, protozoal and miscellaneous infections. In: Ettinger, S.J. (Ed.), *Textbook of veterinary internal medicine*, 1989, 3 ed. Philadelphia: W.B. Saunders, p. 265-297.

Tenter, AM.; Heckeroth, AR.; Weiss, LM. *Toxoplasma gondii*: from animals to humans. *International Journal for Parasitology*, 2000, v.30, n.12-13, p.1217-1258.

Todd, ECD. Preliminary estimates of costs in foodborne diasease in the United States. *Journal of food protection*, 1989, v. 52, p. 595-601.

Ueno, TEH., Gonçalves, VSP., Heinemann, MB., Dilli, TLB., Akimoto, BM., de Souza, SLP., Gennari, SM., Soares, RM. Prevalence of *Toxoplasma gondii* and Neospora caninum infections in sheep from Federal District, central region of Brazil. *Tropical Animal Health and Production*, 2009, v. 41, p. 547–552.

Watson WA., Beverley, JKA. Ovine abortion due to experimental toxoplasmosis. *Vet. Rec.*, 1971, v.88, p.42-45.

Williams, RH., Morley, EK., Hughes, JM., Duncanson, P., Terry, RS., Smith, JE., Hide, G. High levels of congenital transmission of *Toxoplasma gondii* in longitudinal and cross-sectional studies on sheep farms provides evidence of vertical transmissionin ovine hosts. *Parasitology*, 2005, v. 130, p. 1–7.

Wong, SY., Remington, JS. Biology of *Toxoplasma gondii.* Aids, 1993, v. 07, n.03, p.299-316.

Wyss, R.; Sager, H.; Muller, N. et al. Distribution of *Toxoplasma gondii* and Neospora caninum under aspects of meat hygiene. *Schweizer Archiv fur Tierheilkunde*, 2000, v.142, n.3, p.95-108.

Yadav, V.; Chu, NGK; Rais, RH.; Al Safarjalani, ON.; Guarcello, V.; Naguib, FNM.; El Kouni, MH. Synthesis, biological activity and molecular modeling of 6-benzylthioinosine analogues as subversive substrates of *Toxoplasma gondii* adenosine kinase. *Journal of Medicinal Chemistry*, 2004, v. 47, n. 8, p. 1987-1996.

In: *Toxoplasma Gondii*
Editors: M. Goodson and D. Winslow

ISBN: 978-1-62618-349-0


*Chapter III*

# Epidemiology and Control of Toxoplasmosis in Dogs

***Katia Denise Saraiva Bresciani[1,2*], André Luiz Baptista Galvão[2], Amanda Leal de Vasconcellos[2], Italmar Teodorico Navarro[3], Jancarlo Ferreira Gomes[4], Rodrigo Rabelo dos Santos[2], Thaís Rabelo dos Santos[1,2] and Alvimar José da Costa[2]***

[1]Faculdade de Medicina Veterinária de Araçatuba, UNESP, Araçatuba, São Paulo, Brazil
[2]Faculdade de Ciências Agrárias e Veterinárias, Jaboticabal, SP, Brazil
[3]Universidade de Londrina, Paraná, Brazil
[4]Instituto de Biologia da Universidade de Campinas, SP, Brazil

## Abstract

Considering the close proximity between men and dogs and the zoonotic potential of these diseases, determining the occurrence of endoparasitosis, mainly toxoplasmosis, in pets has become increasingly

[*] E-mail: bresciani@fmva.unesp.br.

relevant. This disease is frequently diagnosed in spite of the existent therapeutic and prophylactic measures. Studies of animal toxoplasmosis are essential in view of its transmission to men and its pathogenicity in pets. The dog can be involved in the mechanical transmission of the infection by *Toxoplasma gondii* and in spite of not being the definitive host, it has an epidemiological role in this disease. The high rate of dogs naturally infected by *T. gondii* and the great possibility of adquiring this protozoan, justify the importance of attempting against for the occurrence of this illness. The aim of this chapter wil be clear some clinical aspects, address the participation of small animals in the spread of the disease, as well as discuss methods of diagnosis, therapeutic measures, prophylaxis and control of this disease.

## Introduction

The dog can be involved in the mechanical transmission of the infection by *Toxoplasma gondii* and in spite of not being the definitive host, it has an epidemiological role in this disease (Frenkel and Parker, 1996, Lindsay et al., 1997, Schares et al., 2005).

In humans, toxoplasmosis has been correlated with behavioral changes (Lafferty, 2005) and neurological changes (Pradhan et al., 2007), in episodes of reactivation of the referred disease (Hung et al., 2007).

This disease is important in public health, by causing abortion and congenital abnormalities in several species of intermediate hosts (Tenter et al., 2000). In women, it causes reproductive disorders and sequelae in postnatal life (Bachmeyer et al., 2006, Hung et al., 2007).

Thus, it becomes necessary to elucidate the role of the dogs in this disease dissemination, clarify clinical aspects, discuss diagnosis methods and recommend therapeutic and control actions for this disease.

## Toxoplasmosis Canine

The etiology of toxoplasmosis was elucidated, almost simultaneously, by Nicolle and Manceaux (1908) in Tunisia, North Africa, and by Splendore (1908), in São Paulo, Brazil. Canine toxoplasmosis was first described in Italy by Mello (1910) and in Brazil by Carini (1911) and Carini and Maciel (1913).

Obligate intracellular coccidian protozoan, *T. gondii* (Bonametti et al., 1997), belong to the Kingdom Protista, Sub-kingdom Protozoa and Phylum Apicomplexa (Kawazoe, 2002).

The occurrence of infection by this parasite, in Brazil, is shown in Table 1.

**Table 1. Occurrence of *Toxoplasma gondii* infection in dogs from 1981 to 2012, Brazil**

| Year | Reference | State | Serological tests | Seroprevalence (%) |
|---|---|---|---|---|
| 1981 | Ishizuka et al. | São Paulo | IFAT | 63.80 |
| 1985 | Salata et al. | São Paulo | IFAT | 63.80 |
| 1985 | Germano et al. | São Paulo | IFAT | 91.00 |
| 1992 | Guimarães et al. | Minas Gerais | IFAT | 47.30 |
| 1992 | Freire et al. | Paraná | IFAT | 75.90 |
| 1997 | Navarro et al. | Paraná | IFAT | 23.40 |
| 1998 | Domingues et al. | São Paulo | IFAT e ELISA | 80.00 |
| 1998 | Cabral et al. | Minas Gerais | ELISA | 52.70 |
| 1999 | Garcia et al. | Paraná | IFAT | 84.10 |
| 2003 | Coelho et al. | Rio de Janeiro | IFAT | 46.60 |
| 2003 | Souza et al. | Paraná e São Paulo | MAT | 21.30 |
| 2004 | Cánon-Franco et al. | Rondônia | IFAT | 76.40 |
| 2004 | Reis et al. | Paraná | IFAT | 45.73 |
| 2004 | Ragozo et al. | Paraíba | IFAT | 45.10 |
| 2004 | Rosado et al. | Minas Gerais | IFAT | 62.20 |
| 2004 | Mineo et al. | Minas Gerais | ELISA | 30.30 |
| 2005 | Azevedo et al. | Paraíba | IFAT | 45.10 |
| 2007 | Bresciani et al. | São Paulo | IFAT | 23.10 |
| 2009 | Bresciani et al. | Santa Catarina | IFAT | 22.30 |
| 2009 | Marques et al. | Mato Grosso do Sul | IFAT | 47.60 |
| 2009 | Santos et al. | Mato Grosso | IFAT | 88.50 |
| 2011 | Lopes et al. | Piauí | IFAT | 18.00 |
| 2011 | Araújo | Paraná | MAT | 54.80 |
| 2012 | Langoni et al. | São Paulo | IFAT | 22.00 |

*IFAT: indirect fluorescent antibody test; ELISA: Enzyme-linked immunosorbent assay; MAT: modified agglutination test.

The complete life cycle is heteroxenic, being definitive hosts the cats and intermediate mammals (including humans), birds, reptiles, amphibians and fish. In cats, either the enteroepithelial (with schizogony, merogonia, gametogonia and formation of oocysts) as the extra-intestinal cycle can occur. In intermediate hosts only the asexual cycle occurs (Miller and Frenkel, 1972, Frenkel and Bermudez, 2006).

Oocysts are shed in the feces of domestic or wild cats and by sporogony, become infective, being highly resistant to environmental conditions (Miller and Frenkel, 1972). The protozoan presents itself in more two evolutive forms: (1) tachyzoites, which are structures of rapid multiplication, present in the body fluids during the acute phase of the disease and (2) the bradyzoites, contained within the cyst, present in nervous system and muscles, during the chronic phase of the disease (Lappin, 2005, Gademartori, 2007).

Ingestion of tissue cysts in meat products, consumption of food and water contaminated with sporulated oocysts and congenital infection by tachyzoites represent the three major forms of *T. gondii* transmission (Dubey, 1986).

Other transmission routes, with less epidemiological importance, include the ingestion of contaminated milk, transfer of organic fluids, organ transplantation (Dubey et al. 1990, Powell et al., 2001, Miro et al., 2004) and mechanical transmitting of this zoonosis by dog, which possibly carries oocysts in their pelage (Frenkel and Parker, 1996, Lindsay et al., 1997, Schares et al., 2005).

Antineoplastic therapy and use of glucocorticoids can trigger immunosuppression, reactivation of brain cysts and development of clinical toxoplasmosis in infected hosts (Dubey and Lappin, 2006).

Disease of high prevalence, chronic character, with increased occurrence with age, being more frequent in wandering dogs when compared to domiciled (Souza et al., 2003, Cánon-Franco et al., 2004) kept in cemented environment (Bresciani et al., 2007).

Interesting to note is that after the primary experimental infection of oocysts and tachyzoites of *T. gondii* in *female* dogs, this parasite was detected in samples of saliva, milk and urine by bioassay in mice (Bresciani et al., 2001). Also, in a study of toxoplasmic reinfection during pregnancy, the products of these mothers were born apparently healthy but seropositive (Bresciani et al., 2009).

Arantes et al., (2009) described the presence of *T. gondii* in semen, testes and epididymis in male dogs with experimental toxoplasmosis. Seven days after inoculation of seminal samples positive for the parasite, female dogs with negative titration showed serologic conversion.

Fetal resorption was also evidenced during pregnancy and toxoplasmic cysts were found in the encephalon of pups, confirming the sexual transmission of this infection.

## Pathogenesis and Clinical Signs

In infections resulting from ingestion of tissue cysts or oocysts, occurs inappetence, dyspnea, cough, diarrhea, vomiting, haematemesis, splenomegaly and occasionally hepatomegaly, these manifestations occur by the action of tachyzoites, especially in the digestive system, respiratory system and associated lymphoid organs (Smart et al. 1973).

The retinochoroiditis is a frequent injury in congenital toxoplasmosis. The retina becomes inflamed and necrotic and the pigmented layer breaks by infiltration of inflammatory cells; finally the granulation tissue is formed, invading the vitreous humor (Urqhart et al., 1990).

Eye compromise promoted by *T. gondii* is characterized by its tropism for neural tissue, such as retina. Necrosis with neovascularization of subretinal space and choroid, as well as local inflammatory process involving the retina and its pigment epithelium are damages caused by the tachyzoites (Gazzinelli et al., 1994).

In experimental infections of *T. gondii* in dogs, Abreu et al., (2001), observed an increase of lymph nodes and ophthalmologic abnormalities. Bresciani et al., (1999; 2001) found fever, runny nose, tearing, prostration, lymphadenopathy, premature birth, miscarriage and stillbirth in pregnant dogs. The lymphadenopathy is reported as the most common clinical manifestation in this animal species (Smart et al., 1973, Dubey and Lappin, 2006).

D'Ark Moretti et al., (2002) found widespread clinical signs, especially with neuromuscular impairment, most often in puppies. *T. gondii* muscle injury determine hyperesthesia to palpation, stiff gait and lameness. Arrhythmias and heart failure may also be developed due to myocardial involvement (Lappin, 2005, Dubey and Lappin, 2006).

Neurological signs depend on the location of the parasite in the brain, cerebellum and spinal cord. The proliferation of agents at these sites leads to seizures, cranial nerve deficits, ataxia, tremors, paralysis and paresis (Dubey and Lappin, 2006).

Changes in behavior eventually may occur with the development of the disease, such as apathy or aggression. However, the signs are predominant paresis and paralysis of hind limbs (Corrêa and Corrêa, 1992, Plugge et al., 2011). In naturally acquired toxoplasmosis are the predominant spastic paraparesis, flaccid paraplegia, spastic paraplegia, ataxia with intention tremor, hemiparesis, injuries resulting from the impairment of spinal nerves, spinal cord, cerebellum and unilateral cortex (Averill and Lahunta, 1971, Nesbit et al., 1981, Suter et al., 1984, Plugge et al., 2011).

Silva et al., (2007) considers common the finding of pneumonic and neurologic *manifestations* in this disease. The coccidian *surpass* the blood-brain barrier, intracellularly, affecting endothelial cells, especially capillaries, promoting vasculitis and edema, which can progress to hemorrhage and necrosis (McGavin and Zachary, 2012). Pulmonary lesions are varied, multifocal, interstitial necrosis and proliferation of type II pneumocytes and infiltration of macrophages and neutrophils (McGavin and Zacahry, 2012).

Hoffman et al., (2012) described two cases of toxoplasmosis with skin clinical manifestation in dogs that underwent immunosuppressive treatments, which supposes that the reactivation of latent *T. gondii* infection occurred. The lesions were characterized by multiple nodules alopecic ulcerated or not, with purulent material with generalized pustular dermatitis.

In this disease, are not known clinicopathologic or specific pathognomonic findings, constituting so, a clinical challenge to the veterinarian, especially when involving the nervous system and the vision organ.

For the small animal clinical, are considered, as diagnostics resources, eye and routine laboratory exams, which may suggest the local action and/or systemic of the protozoan.

In biochemical tests, may be evidenced hypoproteinemia, hypoglobulinemia, increased serum enzymatic activities of alanine amino transferase (ALT), alkaline phosphatase (ALP) and an increase in serum levels of bilirubin (Corrêa and Corrêa, 1992). The muscle damages may be associated with the increase of serum enzyme activity of creatine kinase (Dubey et al., 1998).

In neuropathies, the increase of protein concentration, and/or the presence of eosinophils, can be demonstrated in the analysis of cerebrospinal fluid and suggests migration of the parasite to the nervous system, being indicated a better diagnostic investigation for *T. gondii* (Sarmento et al. 2000).

D'Ark Moretti et al., (2002) highlighted the need to include the differential diagnosis of toxoplasmosis in dogs with respiratory symptoms and/or nervous, in relation to other diseases that affect these systems, such as distemper.

Clinical toxoplasmoses are more common in young dogs and commonly is associated with canine distemper virus or Ehrlichia canis. Thus differential diagnosis is important to guide small animal clinical (Dubey, 1993, Baneth et al., 1993).

The canine neosporosis shows the similar clinical feature of toxoplasmosis, with paralysis of the limbs, especially the later ones with hyperextension and can also affect members usually 3-6 weeks after birth and not all simultaneously. As the clinical signs of neosporosis are similar toxoplasmosis, serology, DNA detection by PCR and IHC staining may help in differential diagnosis (Dubey, 2010).

In acute toxoplasmosis, in blood cell count, may be nonregenerative anemia, leukocytosis with neutrophilia, lymphocytosis, monocytosis and eosinophilia (Dubey and Lappin, 2006).

The use of diagnostic imaging methods helps the understanding of the real compromise of local or systemic tissue promoted by the parasite, being important to justify the clinical manifestations demonstrated by the patient.

The detection of *T. gondii* antibodies, by indirect immunofluorescence, as well as by direct agglutination test, hemagglutination test and ELISA are recommended methods for diagnosis and epidemiological survey of this zoonosis (Zhang et al., 2001, Dubey and Lappin, 2006).

The bioassay represents a great tool for the detection of this parasite (Abreu et al., 2002). Histopathological examination, associated to the immunohistochemistry technique, combined with the polymerase chain reaction (PCR) has been widely used in the diagnosis of this infection (Schatzberg et al., 2003, Brresciani et al., 2009).

MacPherson et al., (1993) standardized a PCR test with high sensitivity and specificity based on gene sequences of 18 S Rrna of *T. gondii* for the diagnosis of toxoplasmosis dogs and cats. Another test used is the real time PCR or RT-PCR (real time PCR), the advantage of this test is to quantify the amount of parasites in the sample. (Homan et al., 2000, Piergili et al., 2004).

Drugs such as pyrimethamine, trimethoprim + sulfonamide and doxycycline, doxycycline, azithromycin, minocycline and clarithromycin may be used (Lappin et al., 2004).

Doxycycline and minoclina demonstrated efficacy in experimental studies realized *in vitro* and *in vivo* in mice and humans, in relation to the compromise of the central nervous system (Dubey and Lappin, 2006).

Papich (2009) recommends the combination sulfadiazine-trimethoprim and the trimethoprim-sulfametaxol combination in the therapy and control of

*T. gondii* infection, which is in line with our experience in small animal clinical routine.

The Ponazuril is a compound used in the treatment of coccidiosis in horses, and although its use is not approved in dogs, Swinger et al., (2009) have shown that this compound can be effective in the treatment of toxoplasmosis, suggesting further studies about this drug in small animals.

Barbosa et al., (2012), in studies *in vitro* and *in vivo,* found that enrofloxacin, which proved to be effective with few side effects, can be an interesting alternative in the treatment of clinical manifestation of toxoplasmosis.

Studies are being developed with different drugs to obtain new alternatives therapies for toxoplasmosis, however, currently, hydrochloride clindamycin, sulfadiazine-trimethoprim and trimethoprim-sulfametaxol have satisfactory effects in treatment and control of toxoplasmosis in small animals.

Interesting to consider regarding drugs against *Toxoplasma*: continuous and prolonged use of pyrimethamine or associated with other sulfa cause hematological complications. Such drugs act synergistically in inhibiting the sequential steps in the biosynthesis of folinic acid which is required T.gondii. Thus must be added folinic acid to prevent hematological complications occasionally seen with treatment with pyrimethamine (Diniz, 2003, Mitisuka et al., 2010).

# Toxoplasmosis Canine Control

The control of the oocysts shedding in cats reduces considerably the transmission of the disease to humans and animals, since the sporulated oocysts survive for extended periods in the environment (Mitisuka et al., 2010).

Daily cleaning and the adequate removal of feces should be performed, once the oocysts require at least 24 hours to sporulate and become infectious. It is recommended to avoid the habit of coprophagia observed in some animals, as well as the control of insects and rodents in the environment.

Do not consume and do not provide to the animals raw meat or raw milk from cow, sheep and goat, being indicated the use of commercial products or foods available (Monteiro, 2010).

Personal hygiene behavior must be strictly followed and water from lakes, rivers and wells that do not undergo a pretreatment should not be consumed or provided to animals. The access of dogs and cats to water reservoirs should be prohibited, as well as to dairy and refrigerators (Monteiro, 2010).

From the data obtained, in the studies conducted by our research group, we can infer that there is a need to implement a program of community education directed to the improvement of the basic concepts of control and prevention of canine toxoplasmosis (Tomé et al., 2005, Lima et al., 2008).

Educational programs directed at reducing environmental contamination with *T. gondii* would eventually lower the cost of treating humans for clinical toxoplasmosis (Santos et al., 2010, Silva et al., 2010).

# Conclusion

The occurrence of canine toxoplasmosis is very high, it´s a important zoonosis and there is a need to implement a program of community education directed to reduce the environmental contamination with *T. gondii* and improve of the basic concepts of control and prevention of this disease.

# References

Abreu, C. B. Navarro, I. T.; Salarin, M. R. S.; Bracarense, A. P. F. R. L.; Marana, E. R. M.; Trapp, SM.; Fuginaka, CA.; Prudêncio, LB.; Matos, MR.; Tsutsui, VS. 2001. Aspectos Clínicos, Patológicos e Sorológicos da Toxoplasmose Experimental em Cães Jovens, *Semina: Ciências Agrárias*, Londrina, Vol. 22, N°. 2, 2001, pp. 123-130, 1679-0359.

Abreu, CB.; Navarro, IT.; Reis, AC. F.; Souza, MSB.; Machado, R.; Marana, ERM.; Prudêncio, L B.; Mattos, MR.; Tsutsui, VS. 2002. Toxoplasmose ocular em cães jovens inoculados com *Toxoplasma gondii*. *Ciência Rural*, Santa Maria, Vol. 32, N°. 5, 2002, pp. 807-812, 0103-8478.

Arantes, TP.; Lopes, WD.; Ferreira, RM.; Pieroni, JS.; Pinto, VM.; Sakamoto, CA.; Costa, AJ. 2009. *Toxoplasma gondii*: evidence for the transmission by sêmen in dogs. *Experimental Parasitology*, San Diego, Vol. 123, N°. 2, 2009, pp. 190-194, 0014-4894.

Araújo, DA.; Silva, AV.; Zanette, DF.; Silva, DR.; Correa, NAB.; Velasquez, LG.; Neto, AP. Investigação dos fatores associados à infecção pelo *T. gondii* em cães e seres humanos de Porto Figueira (PR). 2011. *Revista da*

*Faculdade de Medicina Veterinária e Zooctecnia de Botucatu*, Botucatu, Vol.18, N°. 1, 2011, pp. 98-111, 0102-5716.

Averill, DR.; Lahunta, A. 1971. Toxoplasmosis of The Canine Nervous System: Clinicopathological Findings In Four Cases. *Journal American Veterinary Medical Association*, Schaumburg, Vol. 159, N°. 9, 1971, pp. 11-34, 0003-1488.

Azevedo, SS.; Batista, CSA.; Vasconcellos, SA.; Aguiar, DM.; Ragozo, AMA.; Rodrigues, AA. R.; Alves, CJ.; Gennari, SM. 2005. Seroepidemiology of *Toxoplasma gondii* and *Neospora caninum* in dogs from the state of Paraíba, Northeast region of Brazil. *Research in Veterinary Science,* London, Vol.79, N°. 1, 2005, pp.51- 56, 1297-9716.

Bachmeyer, C.; Mouchnino, G.; Thulliez, P.; Blum, L. 2006. Congenital Toxoplasmosis From An Hiv-Infected Woman As A Result Of Reactivation. *The Journal of Infection*, London, Vol.52, N°. 2, 2006, pp.55–57, 0163-4453.

Baneth, G.; Shkap V.; Savitsky I.; Pipano E. 1996. The prevalence of antibodies to *Toxoplasma gondii* in dogs in Israel. *Journal Veterinary Medicine*, Israel, Vol.51, N°. 1, 1996, pp.31-33, 0334-9152.

Barbosa, BF.; Gomes, AO.; Ferro, EA.; Napolitano, DR.; Mineo, JR.; Silva, NM. 2012. Enrofloxacin in able to control *Toxoplasma gondii* infection in both in vitro and in vivo experimental models. *Veterinary Parasitology*, Amsterdam , Vol. 187, N°. 1-2, 2012, pp. 44-52, 0304-4017.

Bonametti, AM.; Passos, JN.; Da Silva, EMK.; Bortoliero, AL. 1997. Surto de Toxoplasmose aguda transmitida através da ingestão de carne crua de gado ovino. *Revista da Sociedade Brasileira de Medicina Tropical*, Uberaba, Vol. 30, N°. 1, p. 21-25, 0037-8682.

Bresciani, KDS.; Costa, AJ. Toniollo, GH.; Sabatini, GA.; Moraes, FR.; Paulillios, AC; Ferraudo, AS. 1999. Experimental toxoplasmosis in pregnant biches. *Veterinary Parasitology*, Amsterdam, Vol. 86, N°. 2, 1999. pp. 143-145, 0304-4017.

Bresciani, KDS.; Toniollo, GH.; Costa, AJC.; Sabatini, GA.; Moraes, FR. 2001. Clinical, parasitological and obstetric observations in pregnant bitches with experimental toxoplasmosis. *Ciência Rural*, Santa Maria, Vol. 31, N°. 6, 2001, pp. 1039-1043, 0103-8478.

Bresciani, KDS.; Costa, AJ.; Nunes, CM.; Serrano, ACM.; Moura, AB.; Stobbe, NS.; Perri, SHV.; Dias, RA.; GENNARI, SM. 2007. Ocorrência de anticorpos contra *Neospora caninum* e *Toxoplasma gondii* e estudo de fatores de risco em cães de Araçatuba (SP). *ARS Veterinária*, Jaboticabal, Vol. 23, N°. 1, 2007, pp. 40-46, 0102-6380.

Bresciani, KDS.; Souza, AP.; Sartor, AA.; Bellato, V.; Teixeira, EB.; Pisetta, GM. Junior, HA. 2009. Ocorrência de anticorpos e fatores de risco para infecção por *Toxoplasma gondii* em cães, nas cidades de Lages e Balneário Camboriú, Santa Catarina, Brasil. *Revista Brasileira de Parasitologia Veterinária*, Jaboticabal, Vol. 18, N°. 3, 2009, pp. 52-56, 1984-2961.

Cabral, DD.; Silva, DAO.; Mineo, JR.; Ferreira, FA.; Duran, FP. 1998. Frequency of anti– *Toxoplasma gondii* antibodies in apparently healthy dogs of the city of Uberlândia – MG. *Revista Brasileira de Parasitologia Veterinária*, São Paulo, Vol.7, N°. 2, 1998, pp.87-90, 1984-2961.

Cademartori, BG. 2007. *Toxoplasmose: perfil sorológico em gestantes atendidas em Postos de Saúde do município de Pelotas-RS*. Universidade Federal de Pelotas - UFPEL: IB, 2007.102p. Dissertação (Mestrado). Programa de Pós-Graduação em Parasitologia.

Cánon-Franco, WA.; Bergamaschi, DP.; Labruna, MB.; Camargo, LM.; Silva, JC.; Pinter, A.; Gennari, SM. 2004. Ocurrence of anti-*Toxoplasma gondii* antibodies in dogs in the urban área of Monte Negro, Rondônia, Brazil. *Veterinary Research*, London, Vol. 28, N°. 2. 2004, pp. 113-118, 1746-6148.

Carini, A. 1911. Infection spontanée du pigeon et du chien due au toxoplasma cuniculi. *Bulletin de la Societe de Pathologie Exotique,* Paris, Vol.4, 1911, pp.518-519, 1961-9049.

Carini, A.; Maciel, J. 1913. Toxoplasmose naturelle du chien. *Bulletin de la Societe de Pathologie Exotique,* Paris, Vol.6, 1913, pp.681-683, 1961-9049.

Coelho, RA.; Kobayashi, M.; Carvalho Junior, LB. 2003. Prevalence of IgG antibodies specific to *Toxoplasma gondii* among blood donors in Recife, Northeast Brazil. *Revista do Instituto de Medicina Tropical de São Paulo*, São Paulo, Vol.45, N. 4, 2003, pp.229-231, 0036-4665.

Corrêa, W. M.; Corrêa, C. N. M. 1992. Toxoplasmose. In:________. *Enfermidades Infecciosas dos Mamíferos Domésticos*. Rio de Janeiro: Medsi, 1992. p. 757-766.

D´Arc-Moretti, L.; Ueno, TE.; Ribeiro, MG.; Aguiar, DM.; Paes, AC.; Pezerico, SB.; Silva, A. V. 2002. Toxoplasmose em cães co-infectados com o vírus da cinomose. *Semina: Ciências Agrárias*, Londrina, Vol. 23, N°. 1, 2002, pp. 85-91, 1679-0359.

Diniz, EMA.; Vaz, FAC. 2003. Qual é a recomendação atual para o tratamento da toxoplasmose congênita?. *Revista da Associação de Medicina Brasileira*, São Paulo, Vol.48, N. 1, 2003, pp.1-10.

Domingues, LM.; Machado, RZ.; Costa, MT.; Carvalho, CS.; Costa, AJ.; Malheiros, EB. 1998. Canine toxoplasmosis: a comparative evaluation of the detection of anti-*Toxoplasma gondii* antibodies by the indirect immunoenzymatic assay (ELISA) and the indirect immunofluorescence reaction (IIF). *Revista Brasileira de Parasitologia Veterinária*, São Paulo, Vol.7, N°. 2, 1998, pp.79- 85, 1984-2961.

Dubey, JP. 1986. Toxoplasmosis. *Journal of the American Hospital Association*, Schaumburg, Vol. 189, N°. 2, 1986, pp. 166-170, 1547-3317.

Dubey, JP.; Greene, CE.; Lappin, MR. 1990. Toxoplasmosis and neosporosis. In: Greene, CE. Infectious diseases of the dog and cat. Philadelphia. W. B. Saunders, p. 818-834, 1990.

Dubey, JP.; Carpenter, JL. 1993. Histologically confirmed clinical toxoplasmosis in cats: 100 cases (1952-1990). *Journal of the American Hospital Association*, Schaumburg, Vol. 203, N°. 11, 1993, pp. 1556-1566, 1547-3317.

Dubey, JP. 1993. *Toxoplasma, Neospora, Sarcocystis* and other tissue cyst-forming of human and animals. In: Krier, JP, editor. Parasitic Protozoa. San Diego: Academic Press; 1993. p. 1-157.

Dubey, JP.; Lappin, MR. 1998. Toxoplasmosis and neosporosis. In: Greene, CE. *Infectious diseases of the dog and cat.* 2.ed. Philadelphia: WB Saunders Company, 1998. Cap. 90, p. 493-503.

Dubey, JP.; Lappin, MR. 2006. Toxoplasmosis and neosporosis. In: Greene CE. *Infectious diseases of the dog and cat.* 3. ed. St. Louis: Elsevier, 2006, p. 754-775.

Dubey, JP. 2010. Toxoplasmosis of animals and humans / Jitender Prakask Dubey. 2nd ed. 2010 by CRC Press Taylor and Francis Group, LLC. London, NewYork. p. 319.

Freire, RL.; Navarro, IT.; Vidotto, O.; Tudury, EA.; Vianna, CC. 1992. Prevalência de Anticorpos anti-*Toxoplasma gondii* em cães atendidos no Hospital Veterinário da UEL-PR. *Semina: Ciências Agrárias*, Londrina, Vol. 13, N°. 1, 1992, pp. 66-69, 1679-0359.

Frenkel, JK; Bermudez, JEV. 2006. Toxoplasmose. In: Veronesi, R.; Focaccia, R. *Tratado de Infectologia*. São Paulo: Atheneu, 2006, p. 1634-1649.

Frenkel, JK.; Parker, BB. 1996. An apparent role of dogs in the transmission of *Toxoplasma gondii* . The probable importance of xenosmophilia. *Annals of the New York. Academy of Sciences*, New York, Vol.791, 1996, pp.402-407, 1749-6632.

Garcia, JL.; Navarro, IT.; Ogawa, L.; Oliveira, RC. 1999. Soroepidemiologia da toxoplasmose em gatos e cães de propriedades rurais do Município de

Jaguapitã, Estado do Paraná, Brasil. *Ciência Rural,* Santa Maria, Vol.29, N. 1, 1999. pp.99-104, 0103-8478.

Garcia JL.; Gennari SM.; Machado RZ.; Navarro IT. 2006. *Toxoplasma gondii*: Detection by mouse bioassay, histopathologic and polimerase chain reaction (PCR) in tissue samples from experimentally infected pigs. *Experimental Parasitology,* New York, Vol. 113, N°. 4, 2006, pp. 267-271, 0014-4894.

Gazzinelli, RT.; Brezin, A.; Nussenblatt, RB.; Chan, CC. 1994. *Toxoplasma gondii*: acquired ocular toxoplasmosis in the murine, protective role of TNF-alpha and IFN-gamma. *Experimental Parasitology*, San Diego, Vol. 78, N°. 2, 1994, pp. 217-219, 0014-4894.

Germano, PML.; Erbolato, EB.; Ishizuka, MM. 1985. Estudo sorológico da toxoplasmose canina, pela prova de imunofluorescência indireta, na cidade de Campinas, 1981. *Revista da Faculdade de Medicina Veterinária e Zootecnia da Universidade de São Paulo*, São Paulo, Vol.22, N°. 1, 1985, pp. 53-58, 0102-5716.

Guimarães, AM.; Ribeiro, MFB.; Lima, JD. 1992. Frequência de anticorpos anti-*Toxoplasma gondii* em cães de Belo Horizonte, Minas Gerais. *Arquivo Brasileiro de Medicina Veterinária e Zootecnia,* Belo Horizonte, Vol.44, N°. 1, 1992, pp.67-68, 0102-0935.

Hoffman, AR.; Cadieu, J.; Kiupel, M.; Lim, A.; Bolin, SR.; Manselli, J. 2012. Cutaneous toxoplasmosis in two dogs. *Journal of Veterinary Diagnostic Investigation*, Columbia, Vol. 24, N°. 3, 2012, pp. 636-640, 1943-4936.

Homan WL.; Vercammen, M; De Braekeleer, J. 2000. Identification of a 200 to 300 fold repetitive 529 bp DNA fragment in *Toxoplasma gondii*, and its use for diagnostic and quantitative PCR. *International Journal of Parasitology*, Oxford, Vol.30, N°. 1, 2000. pp. 69-75, 1383-5769.

Hung, CC.; Fan, CK.; Su, KE.; Sung, FC.; Chiou, HY.; Gil. V.; Ferreira, MDAC.; Carvalho, JM.; Cruz, C.; Lin, YK.; Tseng, LF.; Sao, KY.; Chang, WC.; Lan; HS.; Chou, SH. 2007. Serological screening and toxoplasmosis exposure factors among pregnant women in the Democratic Republic of Sao Tome and Principe. *Transactions of the Royal Society of Tropical Medicine and Hygiene,* Oxford, Vol. 101, N°. 2, 2007, pp. 134-139, 0035-9203.

Ishizuka, MM.; Miguel, O.; Brogliato, DF. 1981. Incidência de infecção por *Toxoplasma gondii* em cães do município de São Paulo. *Revista da Faculdade de Medicina Veterinária e Zootecnia da Universidade de São Paulo,* São Paulo, Vol.18, N°. 2, 1981, pp.161-165, 1413-9596.

Jones, JL.; Dubey, JP. 2012. Foodborne Toxoplasmosis. *Clinical Infectious Diseases*, Vol. 18, 2012. pp. 1-7, 1537-6591.

Kawazoe, U. 2002. *Toxoplasma gondii*. In: Neves D. P. *Parasitologia Humana*. 10 ed. São Paulo: Atheneu, pp. 147-156.

Lappin, MR. et al. 2004. *Segredo em Medicina Interna de Felinos*, Porto Alegre: Armed, 2004.

Lappin, MR. 2005. Protozoal and Miscellaneous Infections. In: Ettinger, S.J.; Feldman, EC. *Textbook of Veterinary Internal Medicine*. 6° ed. St Louis: Elsevier Saunders. 2005, cap. 168, pp. 638-646.

Lafferty, KD. 2005. Look what the cat dragged in: do parasites contribute to human cultural diversity? *Behavioural Processes*, Amsterdam, Vol.68, N°. 3, 2005, pp.279-282, 0376-6357.

Langoni, H.; Matteuci, G.; Medici, B.; Camossi, LG.; Richini-Pereira, VB.; Da Silva, RC. 2012. Detection and molecular analysis of *Toxoplasma gondii* and *Neospora caninum* from dogs with neurological disorders. *Revista da Sociedade Brasileira de Medicina Tropical*, Uberaba, Vol. 4, N°. 3, 2012, pp. 365-368, 0037-8682.

Lima, FF.; Koivisto, MB.; Perri, SHV.; Bresciani, KDS. 2008. O conhecimento de idosos sobre parasitoses em instituições não governamentais do Município de Araçatuba, SP. *Revista Ciência em Extensão,* São Paulo, Vol. 4, N°. 1, 2008. pp. 77-85, 1679-4605.

Lindsay, DS.; Dubey, JP.; Butler, JM.; Blagburn, BL. 1997. Mechanical transmission of *Toxoplasma gondii* oocysts by dogs. *Veterinary Parasitology*, Amsterdam, Vol.73, N°. 1/2, 1997. pp.27-33, 0304-4017.

Lopes, MG.; Mendonça, IL.; Fortes, KP.; Amaku, M.; Pena, HFJ.; Gennari, S. M. 2011. Presence of antibodies against *Toxoplasma gondii, Neospora caninum* and *Leishmania infantum* in dogs from Piauí. *Revista Brasileira de Parasitologia Veterinária*, Jaboticabal, Vol. 20, N°. 2, 2011, p.111-114, 1984-2961.

Marques, JM.; Isbrecht, FB.; Lucas, TM.; Guerra, IMP.; Dalmolin, A.; Silva, RC.; Langoni, H.; Silva, A. V. 2009. Detecção de anticorpos anti-*Toxoplasma gondii* em animais de uma comunidade rural do Mato Grosso do Sul, Brasil. *Semina: Ciências Agrárias*, Londrina, Vol. 30, N°. 4, 2009, pp. 889-898, 1679-0359.

Macpherson, JM.; Gajadhar, AA. 1993. Sensitive and Specific Polymerase Chain Reaction Detection of *Toxoplasma gondii* for Veterinary and Medical Diagnosis. *Canadian Journal Veterinary Research*, Ottawa, Vol. 57, N. 1, 1993, pp. 45-48, 0830-9000.

Mcgavin, MD.; ZACHARY, JF. 2012. *Pathologic basis of veterinary disease*, 5 ed. Elsevier: St. Louis, 2012, 1321p.

Mello, V. 1910. Un cas de toxoplasmose du chien observé à Turin. *Bulletin Society Pathology,* Itália, Vol.28, 1910, pp.359-363, 1961-9049.

Nicolle. CS.; Manceaux, L. 1908. Sur une infection à corps de Leishman (on organismes voisins), du gondii. *Comptes Rendus l'Academie des Sciences*, Montrouge, Vol.147, 1908, pp.763-766, 1310-131.

Miller, NL.; Frenkel, JK.; Dubey, JP. 1972. Oral infections with Toxoplasma cysts and oocysts infelines, other mammals and birds. *Journal of Parasitology*, Lawrence, Vol.58, nº.5, 1972, pp.928-937, 1937-2345

Mineo, TWP.; Silva, DAO.; Näslund, K.; Björkman, C.; Uggla, A.; Mineo, JR. 2004. *Toxoplasma gondii* and *Neospora caninum* serological status of different canine populations from Uberlândia – Minas Gerais. *Arquivos Brasileiros de Medicina Veterinária*, Belo Horizonte, Vol. 56, N°. 3, 2004, pp. 414-417, 0102-0935.

Monteiro, S. G. 2010. *Parasitologia na Medicina Veterinária*, São Paulo: Roca, 2010.

Nesbit, JW.; Lourens, DC.; Williams, MC. 1981. Spastic paresis in two littermate pups caused by *Toxoplasma gondii. Journal Soulth Africa Veterinary Association*, Pretoria, Vol. 52, N°. 3, 1981, pp. 243-246, 0895-9811.

Papich, MG. 2009. *Manual Saunders: Terapêutico Veterinário*. 2 ed. São Paulo: MedVet, 2009, 774p.

Nicolle, CS.; Manceaux, L. 1908. Sur une infection à corps de Leishman (on organismes voisins), du gondii. *Comptes Rendus l'Academie des Sciences*, Montrouge, Vol.147, 1908, pp.763-766, 1310-131.

Piergili, F.D. 2004. Problems and limitations of conventional and innovative methods for the diagnosis of Toxoplasmosis in humans and animals. *Parasitologia*, Roma, Vol. 46, N°. 1-2, 2004, pp.177-81, 0048-2951.

Plugge, NF.; Ferreira, FM.; Richartz, RRTB.; Siqueira, A.; Dittich, RL. 2011. Ocurrence of antibodies against Neospora caninum and *Toxoplasma gondii* in dogs with neurological sings. *Revista de Parasitologia Veterinária*, Jaboticabal, Vol. 20, N°. 3, 2011, pp. 202-206, 0103-846X.

Pradhan, S.; Yadav, R.; Mishra, VN. 2007. Toxoplasma meningoencephalitis in HIV-seronegative patients: clinical patterns, imaging features and treatment outcome. *Transactions of the Royal Society of Tropical Medicine and Hygiene*, London, Vol.101, N°. 1, 2007, pp.25-33, 0035-9203.

Ragozo, AMA.; Azevedo, SS.; Vasconcellos, SA.; Batista, CSA.; Aguiar, DM.; Rodrigues, AA. R.; Alves, CJ.; Gennari, SM. 2004. *Toxoplasma gondii* em cães na cidade de Campina Grande, Paraíba: soroepidemiologia e fatores de risco. In: Congresso Brasileiro De Parasitologia Veterinária, 13., 2004, Ouro Preto. *Anais*... Ouro Preto: Colégio Brasileiro de Parasitologia Veterinária, 2004. p.217.

Reis, CR.; Reis, HR.; Gonçalves, DD.; Lopes, FMR.; Carletti, RT.; Silva Filho, MF.; Freire, RL.; Navarro, IT. 2004. Levantamento sorológico da toxoplasmose em cães de Bela Vista do Paraíso – Paraná – Brasil. In: Congresso Brasileiro De Parasitologia Veterinária, 13., 2004, Ouro Preto. *Anais*...Ouro Preto: Colégio Brasileiro de Parasitologia Veterinária, 2004. p.211.

Rosado, IR.; Guimarães, AM.; Oliveira, TMFS. 2004. Prevalência De Anticorpos Anti *Toxoplasma gondii* em cães atendidos em clínicas veterinárias de Lavras, Minas Gerais. In: Congresso Brasileiro De Parasitologia Veterinária, 13., 2004, Ouro Preto. *Anais*...Ouro Preto: Colégio Brasileiro de Parasitologia Veterinária, 2004. p.217.

Salata, E.; Yoshida, ELA.; Pereira, EA.; Corrêa, FMA. 1985. Toxoplasmose em animais silvestres e domésticos da região de Botucatu, Estado de São Paulo, Brasil. *Revista do Instituto de Medicina Tropical de São Paulo,* São Paulo, Vol.27, N°. 1, 1985, pp.20-22, 0102-5716.

Santos, TR.; Nunes, CM.; Luvizotto, MCR.; Moura, AB.; Lopes, WDZ. ; Costa, AJ.; Bresciani, K DS. 2010. Detection of Toxoplasma gondii oocysts in environmental samples from public schools. *Veterinary Parasitology*, Amsterdam, Vol. 171, N°. 1-2, 2010, pp. 53–57, 0304-4017.

Santos, TR.; Costa, AJ.; Toniollo, GH.; Luvizotto, MCR.; Benetti, AH.; Santos, RR.; Matta, DH.; Lopes, WDZ., Oliveira, JA.; Oliveira, GP., 2009. Prevalence of anti-*Toxoplasma gondii* antibodies in dairy cattle, dogs, and humans from the Jauru micro-region, Mato Grosso state, Brazil. *Veterinary Parasitology*. Vol. 161, 2009, pp. 324–326, 0304-4017.

Sarmento, LVC.; Tudury, E.; Albuquerque, ER.; Magalhães, PKL. 2000. Coleta, análise e interpretaão do líquido céfalo-raquidiano de cães e gatos – revisão. *Clínica Veterinária*, São Paulo, ano Vol. 10, N°. 25, 2000, pp.19-26, 1413-571X.

Schares, G.; Pantchev, N.; Barutzki, D.; Heydorn, AO.; Bauer, C; Conraths, FJ. 2005. Oocysts of *Neospora caninum*, *Hammondia heydorni*, *Toxoplasma gond*ii and *Hammondia hammondi* in faeces collected from dogs in Germany. *International Jounal forParasitology*, Oxford, Vol.35, N°. 14, 2005, pp.1525-1537, 1383-5769.

Schatzberg, SJ.; Haley, NJ.; Barr, SC.; Lahunta, A.; Olby, N.; Munana, K.; Sharp, NJH. 2003. Use of a multiplex polymerase chain reaction assay in the antemortem diagnosis of toxoplasmosis and neosporosis in the central nervous system of cats and dogs. *American Journal of Veterinary Research*, Chicago, Vol.64, N°. 12, 2003, pp.1507-1513, 0002-9645.

Silva, MC.; Fighera, RA.; Brum, JS.; Graça, DL.; Kommer, GD.; Irifoyen, LF.; Barros, CSL. 2007. Aspectos clínicos patológicos de 620 casos neurológicos de cinomose em cães. *Pesquisa Veterinária Brasileira*, Rio de Janeiro, Vol. 27, N°. 5, 2007, pp. 215-220, 0100-736X.

Smart, ME.; Downey, RS.; Stoxkdale, PH. 1973. Toxoplasmosis in cat associated with cholangitis and progressive pancreatitis. *The Canadian Veterinary Journal*, Ottawa, Vol.14, N°. 122, 1973, pp. 313-316, 0830-9000.

Souza, H. J. 2003. Coletâneas em medicina e cirurgia felina, *Coletâneas em medicina e cirurgia felina*. Rio de Janeiro: L. F.. Livros de Veterinária, 2003.

Souza, SLP.; Gennari, SM.; Yai, LEO.; D'auria, SRN.; Cardoso, SMS.; Guimarães-Junior, JS.; Dubey, JP. 2003. Ocurrence of *Toxoplasma gondii* antibodies in sera from dogs of the urban and rural areas from Brazil. *Revista Brasileira de Parasitologia Veterinária*, São Paulo, Vol.12, N°. 1, 2003, pp.1-3, 1984-2961.

Splendore, A. 1908. Un nuovo protozoa parasita dei Gonigli incontrato nelle lesioni anatomiche d'una malattia chericorda in multipunti il Kala-Azar dell Uomo. *Revistada Sociedade Scientifica de São Paulo*, São Paulo, Vol.3, 1908, pp.109-112, 0104-5687.

Suter, MM.; Hauser, B.; Palmer, D. G.; Oettli, P. 1984. Polymyositis – polyradiculitis due to toxoplasmosis in the dog: serology and tissue biopsy as diagnostic aids. *Zentalblatt Veterinarmedizin*, Hamburg, Vol.31, N°. 10, 1984, pp. 792-798, 1439-1450.

Swinger, RL.; Shmidt, KA.; Dubeielzig, RR. 2009. Keratoconjunctivitis associated with *Toxoplasma gondii* in a dog. *Veterinary Ophthalmology*, California, Vol. 12, N. 1, 2009, pp. 56-60, 0004-2749.

Mitsukal-Breganó, R.; Lopes-Mori FMR.; NAVARRO IT. 2010. Toxoplasmose Adquirida na Gestação e Congênita: vigilância em saúde, diagnóstico, tratamento e condutas. *Ed. EDUEL*, Londrina. 2010 pp. 66, 1519-9584.

Tenter, AM.; Heckeroth, AR.; Weiss, L. M. 2000. *Toxoplasma gondii*: from animals to humans. *International Journal for Parasitology*; New York, Vol.30, 2000. pp.1217-1258, 1383-5769.

Tomé, RO.; Serrano, ACM.; Nunes, CM.; Perri, SHV.; Bresciani, KDS. 2005. Inquérito epidemiológico sobre conceitos de zoonoses parasitárias para professores de escolas municipais do ensino infantil de Araçatuba-SP. *Revista Ciência em Extensão,* São Paulo, Vol.2, N°. 1, 2005, pp.38, 1679-4605.

Urquhart, GM.; Amour, J.; Duncan, JL.; Dunn, AM.; Jennings, FW. 1990. *Parasitologia veterinária*, Rio de Janeiro: Guanabara Koogan, 1990, 306p.

Zhang, S.; Wei, MX.; Ding, ZY.; Xu, XP. 2001. Comparison a modified agglutination test (MAT), IHAT and ELISA for detecting antibodies to *Toxoplasma gondii. Acta Parasitologica Medica Entomologica Sinica*, China, Vol. 8, N°. 4, 2001, pp. 199-203, 1896-1851.

In: *Toxoplasma Gondii*
Editors: M. Goodson and D. Winslow
ISBN: 978-1-62618-349-0


*Chapter IV*

# Toxoplasmosis in Bovine

***Thaís Rabelo dos Santos*[1], *Rodrigo Rabelo dos Santos*[1], *Alvimar José da Costa*[1] *and Katia Denise Saraiva Bresciani*[2]**
[1]FCAV, UNESP - Jaboticabal, Jaboticabal, SP, Brazil
[2]Department of Production and Animal Health, FMVA, UNESP, Jaboticabal, SP, Brazil

## Abstract

*Toxoplasma gondii* is a parasite of the phylum *Apicomplexa*, *Coccidia* order, being an obligate intracellular protozoan that affects humans and diverse vertebrate hosts. The genus was named *Toxoplasma* (from the greek *toxon* = arc, *plasma* = shape) referring to its shape of "first quarter" or "half moon" and *gondii* in reference to the rodent where the parasite was isolated. In retrospect, the correct name for the parasite should have been *Toxoplasma gundii*, however Nicolle E Manceaux (1908) incorrectly identified the host as *Ctenodactylus gondi*.

The felines, in particular the domestic cats, are the definitive hosts of the parasite, whereas warm-blooded animals are the intermediate hosts. Toxoplasmosis is one of the most spreaded zoonoses in the whole world, however the frequency of infection started to be characterized from 1948, with Sabin and Feldman, and then with other researchers, developing the first diagnostic of the disease. The toxoplasmosis affects about two million people worldwide. The prevalence of seropositive individuals for toxoplasmosis increases with age and differs depending on the population cultural patterns, dietary habits and urban or rural origin. It is known that in areas of tropical or subtropical humid climate, the prevalence is higher, once this type of weather favors the survival of oocysts in the environment.

It is estimated that about 70% of the brazilian population has already been infected at some point in life. In Brazil, it was found that in every 1,000 live births, five children presented the disease. In different countries, the seroprevalence has been shown between 0% to 90% in the human population. Some countries as Thailand and Japan present low prevalence (<20%).

However, Australia, Poland, UK and Belgium show average prevalence (between 23 and 53%), while Brazil, Tahiti and France present high prevalence (> 60%). The importance of the animal toxoplasmosis occurs, at first, by the fact that infected animals serve as direct or indirect source of the human infection and, secondly, by the various reproductive changes such as abortion, neonatal mortality and congenital defects consequent to the *T. gondii* infection, which represent significant losses in animals of economic interest and company. In the world literature on detection of anti-*T. gondii* in bovines, discrepant values are found, ranging from 0 to 99%. In Brazil, this variation in occurrence values is also observed ranging from 1.03% to 71.00%. Infection rates given for sheep flocks in Brazil, vary from 8% to 55%. The infection of herbivores occurs primarily through ingestion of oocysts present in food and contaminated soils. Infection of herds may be associated with the ingestion of grains contaminated by oocysts and the stabling of animals at sites containing feline manure may increase the risk of infection. Direct contact with felines had little epidemic consequence, although the presence of those infected animals indicates a contaminated environment, incurring risk to the human population and other animals.

## Introduction

*Toxoplasma gondii* is a parasite of the phylum Apicomplexa, Coccidia order, being an obligate intracellular protozoan (Pereira-Bueno et al, 2004; Rorman et al., 2006), which affects humans and many vertebrate hosts (Dubey and Beattie, 1988; Miró et al., 2004; Garcia et al., 2006).

This protozoan was first described in laboratory rabbits by Splendore (1908) in São Paulo, Brazil, and simultaneously by Nicolle and Manceaux (1908) in Tunisia, North of Africa, found in the brain of a wild rodent, the *Ctenodactylus Gundi.*

The genus was named *Toxoplasma* (in Greek toxon = arc, plasma = form) referring to its form as "first quarter" or "half moon", and *gondii* in reference to the rodent where the parasite was isolated (Black and Boothroys, 2000).In retrospect, the correct name for the parasite should have been *Toxoplasma gundii*, however Nicolle and Manceaux (1908) had incorrectly identified the host as *Ctenodactylus gondi* (Dubey, 2007).

The first reports of the disease in humans were conducted in 1920 and 1930.In 1923, Jankü observed the agent in the eye of a 11 months child, in Prague, Czechoslovakia. Torres in 1927, in Rio de Janeiro, described the *T. gondii* as cause of congenital meningoencephalitis, myocarditis and myositis in a newborn with 29 days of life. In both studies the parasite was described as *Encephalitozoon,* a protozoan difficult to distinguish from *T. gondii*, without special stains.

In 1937, Wolf and Cowan, reported the disease in many children and showed transplacental transmission, in addition to the isolation of the parasite by animal inoculation and, in the 40s, Pinkerton and Weinman (1940) and Pinkerton and Henderson (1941) reported the acute toxoplasmosis in adults (Amato Neto et al., 1995).

*T. gondii* natural infection in cattle was first reported in Ohio, USA which also reported the first experimental infection by this parasite in cattle.

The cats, particularly domestic cats, are the definitive hosts of the parasite (Dubey and Beattie, 1988; Lappin, 1993; Aslantas et al. 2005; Hutchison, 1965; Faria et al. 2007) while warm blooded animals are the intermediate hosts (Dubey, 2004, Gauss et al., 2005; Jittapalapong et al., 2005; Garcia et al., 2006; Silva et al., 2007).

# Morphology and Life Cycle of the Parasite

In its life cycle, *T. gondii* has three infectious forms: the sporozoites (oocysts), tachyzoites (individually or in groups) and bradyzoites (tissue cysts) (Miller et al., 1972).

The oocysts, final product of sexual reproduction which occurs only in the digestive tract of cats, their definitive hosts, thrive in the presence of oxygen and at lower temperatures than the body, remaining viable, depending on environmental conditions, for up to five years in the presence of moisture and is very resistant to most disinfectants (Vidotto et al., 1986).

Morphologically, the oocysts have a double wall quite resistant to the envirolment conditions (Dubey et al., 1970; Tenter et al. 2 000) measuring, approximately, 10x13 μm and have, in its interior, two sporocysts, measuring 2x8 μm each.Each sporocyst has four 2x8 μm sporozoites (Dubey et al. 1998).They are the key to the maintenance of the infection, as carnivorism and congenital transmission do not complete the life cycle of the coccidia in intermediate hosts (Neto et al., 1995).

The tachyzoites, measuring approximately 4 x 9 μm in length by 2 x 4 μm in diameter, have different structures common to animal cells, such as mitochondria, endoplasmic reticulum and Golgi apparatus, beside characteristic structures of the phylum, such as polar rings, conoid , rhoptries and micronemes.These rapid multiplication forms spreads into the body of the intermediate hosts by blood and lymph, multiplying asexually inside the cells by repeated endodyogeny (Dubey et al. 1998).Some tachyzoites, after the host cell invasion, grow more slowly forming bradyzoites which will give origin to the cysts. The bradyzoites are morphologically similar to tachyzoites, however multiply slowly express stage-specific molecules and are functionally different (Montoya and Liensenfeld, 2004), since the tissue cysts contain hundreds or thousands of bradyzoites (Dubey et al. 1998).The cysts, measuring approximately 10 to 100 μm in diameter, can develop into various organs such as lung, liver, kidneys, but are more prevalent in muscle (skeletal and cardiac) and nervous tissue, including brain and eyes.

The intact cysts probably cause no harm and may persist for long periods or throughout the life of the host without causing any inflammatory response or arouse significant tissue response (Dubey, 2004).The destination of tissue cysts is not fully known, but it is proposed it that can break during life, releasing bradyzoites that can be destroyed by the immune system or form new cysts (Dubey, 1993).

## Relevance

Toxoplasmosis affects, approximately, two million people worldwide (Lindston et al., 2006).The importance of animal toxoplasmosis arises, first, because infected animals serve as direct or indirect source of infection to humans, and second, the various reproductive alterations such as abortion, neonatal mortality and birth defects consequent to *T. gondii* infection which represent significant losses in animals of economic interest (Sawadogo et al., 2005; Yu et al., 2007) and company (Bresciani et al., 1999).

Toxoplasmosis is a zoonosis of major spread worldwide (Clementino et al., 2007), but the frequency of infection began to be characterized from 1948, with Sabin and Feldman, and then with other researchers, to develop the first diagnostic evidence of the disease.

## Prevalence

Approximately 30% of the human population have antibodies against this protozoan (Pena et al., 2006) and the prevalence is even higher in other parts of Europe, Central America and South America (Dubey, 1996).

In different countries, the seroprevalence is shown between 10% to 90% in the human population (James, 1996; Carruthers, 1999).In Brazil, seroprevalence has been determined between 50% and 80%. Some countries, such as Thailand and Japan, have low prevalence (<20%). However, Australia, Poland, UK and Belgium show average prevalence (between 23 and 53%), while Tahiti and France have high prevalence (> 60 %) (Avelino et al., 2003).

Brazil presents human indexes that are among the highest reported, where recorded serosurveys showed a prevalence ranging from 30.34 to 97.10%.

The prevalence of toxoplasmosis seropositive individuals increases with age and differs depending on the cultural patterns of the population, dietary origin and urban or rural procedure. It is known that in tropical or subtropical humid climate, the prevalence is higher, because this type of climate, which favors the survival of oocysts in the environment (Souza et al., 2002; Guimarães et al., 1999; Amendoeira et al . 2003).

A higher seropositivity is an indicator of greater public exposure to determinants of infection.

In the literature on detection of *anti-T. gondii* in cattle, values are discrepant, with a variation from 0 to 99% of the animals studied positive (Hall et al., 2001).In Brazil, this variation in occurrence values is also observed, ranging from 1.03% (GONDIN et al., 1999) to 49.17% (Costa et al., 2001).

# Infection Sources

## Cattle

The infection of herbivores occurs primarily through ingestion of oocysts in feed and contaminated soils (Dubey, 1986). Infection of livestock can be associated with ingestion of oocysts contaminated grain and stabling of animals at sites contaminated with cats manure may increase the risk of infection (Meireles, 2001).

Cattle may be relatively resistant to infection, since the occurrences of cysts in the muscles is less frequent and persist for shorter time when compared to other animal species (Dubey, 1994).

For Stalheim et al. (1980) *T. gondii* appears to be less infective to cows in gestation when compared with ewes, sows and bitches.

Dubey and Thulliez (1993) using calfs inoculated with *T. gondii* oocysts, found the remaining of viable cysts in muscle up to 1191 days post-infection, but antibody titers remained high for two years, becoming serologically negative after this period.

Despite parasitic isolation be more difficult, it has been obtained from bovine retina and diaphragm and congenital transmission was confirmed by the presence of the parasite in fetuses of pregnant cows (Amato Neto et al., 1995)

Daguer et al. (2004) studied the prevalence of antibodies anti-*T. gondii* in cattle and slaughterhouse workers in northern Paraná, and found that 41.4% of the animals were seropositive (IgG-IFA) and 84.4% of humans showed seropositivity (IgG-ELISA), suggesting that beef can play important role in maintaining toxoplasmosis, by the not hygienic handling of raw meat.

## Humans

Interest in the survey of transmission forms of *T. gondii* increased when high prevalences of around 60% positivity in human populations were not explained only by reduced frequency associated with congenital transmission (Amato Neto et al., 1995).

Human infection results from the ingestion of cysts of bradyzoites in raw or undercooked meat (Garcia et al., 2006) and ingestion of sporulated oocysts from feces of infected dogs and cats (Frenkel and Parker, 1996; Lappin, 2004; Clementino et al. 2007).

*T. gondii* infection can also be acquired by blood transfusion (Damian Figueroa, et al., 1998), organ transplantation (Munir et al., 2000), laboratory accidents (Wong and Remington, 1993) and ingestion of tachyzoites contaminated milk (Chiari et al., 1984).

Mead et al. (1999), found that toxoplasmosis is the third leading cause of foodborne deaths in the United States and estimated that there are approximately 1.5 million new acute infections per year, being 15% asymptomatic.

Authors such Giraldi et al. (2002) and Santos et al. (2010), report that the transmission of *T. gondii* by ingestion of *T. gondii* oocysts is the most common route of infection in Latin America.

In a group of followers of the Seventh-day Adventist in the U.S.A., who do not consume meat in their diet because of the religion, was found a lower risk of infection compared with the general population (Roghmann et al., 1999).

In Europe, the percentage of foodborne toxoplasmic infection is not known, however it was estimated between 30 to 63% (Giessen et al., 2007).

Direct contact with felines presents epidemiological little consequence, however the presence of infected animals may indicate a contaminated environment, bringing risk to the human population and other animals (Santos et al., 2009).

Water has been characterized as the primary source of transmission of *T. gondii* to humans in endemic areas (Aramini et al., 1999; Pena et al., 2007).Between November 2001 and January 2002, Brazil recorded the largest outbreak of toxoplasmosis in the world, in Santa Isabel do Ivaí, PR, due to contamination of water supplies in the city, by oocysts present in the faeces of a female cat. An outbreak was reported in 462 patients with suggestive serology of acute infection and 1255 persons were seropositive for IgG *anti-T. gondii*.In some regions of Brazil this transmission mechanism may be contributing to the high prevalence rates (BRAZIL, 2002).

The milk and dairy products can be a major source of human infection with *T. gondii,* reinforcing the importance of milk pasteurization before any processing or ingestion (Hiramoto et al. 2001).

In immunocompetent individuals, the disease is usually benign, however it is necessary to draw attention to ocular toxoplasmosis, often associated with outbreaks of foodborne or water (Derouin et al., 1995; Roberts and McLeod, 1999).

# Diagnosis

Confirmation of the diagnosis depends on the isolation of the parasite, the demonstration of the organism in histological lesions and positive serodiagnosis.

The serological diagnosis can be performed by showing ascending titration of anti-*T. gondii* in paired sera, or by demonstration of high titer antibody serum in a single serum sample, but no evidence of ascending or high titration do not exclude the diagnosis of toxoplasmosis (Lappin, 2004).

There are several valid serological tests, being the most common, the indirect immunofluorescence assay (IFA), indirect hemagglutination (IHA), complement fixation (CF), Sabin-Feldman test (dye test) and enzyme linked immunosorbent assay (ELISA) (Sikes, 1982).

Ttoxoplasmosis laboratorial diagnosis is of great importance, since the infection, both in humans and in domestic and wild animals, can take clinical conditions that are easily confused with diseases such as, viruses, leptospirosis, brucellosis, chlamydiosis and neosporosis, complicating the adoption of specific actions of treatment and control (Vidotto, 1992; Da Silva et al., 2002).

Sabin and Feldman (1948) developed the coloring test, also known as the Dye Test. Following Fulton and Turk (1959), were the first to develop an agglutination test, which revealed low specificity and the need of a large number of tachyzoites in each test. Subsequently, Desmonts and Remington (1980), improved reproducibility and sensitivity of the method.

There is a wide variation in the prevalence of *T. gondii* infection in different regions of the world, due to the different techniques employed, a fact that makes the results not amenable to comparison (Chhabra et al., 1985).

The indirect immunofluorescence assay (IFA) is the most widely used test for the diagnosis of toxoplasmosis, being used as the gold standard. For this, titles of 16 or higher were considered positive for *T. gondii* (Dubey and Beattie, 1988). Currently are considered to be positive, titers higher or equal to 64 (Costa et al., 1977; Souza, 2001). Uchôa et al. (1999) detected a sensitivity of 83.87% and specificity of 79.16% for the IFA-IgG.

In the acute phase of toxoplasmosis, first there is the production of immunoglobulin M (IgM), followed by the production of immunoglobulin G (IgG).The infection may also produce immunoglobulin A (IgA) if the transmission was by oral via. For immunofluorescence, IgM antibodies may be assayed 1-2 weeks after the onset of infection, reaching a peak at 6 to 8 weeks, whereupon decline. Low titers may persist for more than 12 months. The IgG antibody persists throughout life in most patients (Goldsmith, 1998).

Langoni et al. (1999), investigated anti-*T. gondii* in serum samples from 352 sheep from 18 properties in the state of São Paulo, by IFA and indirect hemagglutination techniques. The IFA revealed 55.1% of positivity, whereas only 30.4% for indirect hemagglutination. These data support the need for standardization of serological techniques used in the diagnosis of this infection.

The direct agglutination method (DAM) has been used to demonstrate agglutinins anti-*T. gondii* in various species of domestic and wild animals (Da Silva et al., 2002). This is a simple test, requires no reagent species-specific or sophisticated devices, such as immunofluorescence microscopy, and may be used in samples of human sera as well in different animal species.

Caporali et al. (2005), investigated, by IFA and MAD, the presence of anti-[i]*T. gondii* in serum of pigs from Botucatu region and from the state of Pernambuco, with 757 sera tested by IFA, which 2.11% (16) were positive; MAD were analyzed in 759 sera, with 1.32% (n = 10) of seropositives.

Minho et al. (2004) evaluating the MAD and IFA for detecting anti-[i]*T. gondii* in sera from 46 experimentally infected pigs, obtained a sensitivity of 95.7% and specificity of 97.8% for both methods.

# Prophylaxis

The cysts are destroyed after freezing at - 20 ° C or heating above 65 ° C (Dubey et al., 1986; Amato Neto et al., 1995) thus forming, freezing and full cooking of meat, some forms of transmission prophylaxis of human toxoplasmosis (Tenter et al., 2000).

Safe and effective vaccines against human toxoplasmosis to prevent congenital infection or reactivation of cysts have not yet been developed. Consequently, the best way to avoid the effects of the disease is through prevention (Jones et al., 2001).

Hygienic-dietary guidelines are very important on how to avoid toxoplasmosis, especially in caring for cats, proper cooking of meat, intake of fresh cheese and pasteurized milk (Hiramoto et al., 2001), the treatment of water and washing fruits and vegetables (Remington et al., 1995).

# Conclusion

Clinical toxoplasmosis in cattle is rare and abortions are uncommon. Many reports of bovine abortion due to *T. gondii* are actually due to *N. caninum*. Attempts to isolate *T. gondii* from seropositive cattle are often unsuccessful, indicating that beef may not be a significant source of human infection (Dubey et al., 2005). For example, no *T. gondii* was isolated from samples of beef obtained from retail markets in the United States (Dubey et al., 2005). However, viable tissue cysts can remain in cattle for up to 1191 days (Dubey and Thulliez, 1993). Additional studies are needed to document these experimental findings fully.

# References

Amato Neto, V.; Medeiros, Eas.; Levi, Gc.; Duarte, Mis. 1995. *Toxoplasmose*, 4.ed. São Paulo, Savier, 1995, 154p.

Amendoeira, Mrr.; Sobal, Caq.; Teva, A.; De Lima, Jn.; Klein, CH. Inquérito sorológico para a infecção por *Toxoplasma gondii* em ameríndios isolados, Mato grosso. *Ver. Soc. Bras. Med. Trop.,* 2003, v.36, n.6, p.671-676.

Aramini, Jj.; Stephen, C.; Dubey, Jp.; Engelstoft, C.; Schwantje, H.; Ribble, Cs. Potential contamination of drinking water with *Toxoplasma gondii* oocysts. *Epidemiology and Infection*, 1999, v.122, n.02, p.305-315.

Aslantas, Ö.; Özdemir, V.; Kilic, S.; Babür, C. Seroepidemiology of leptospirosis, toxoplamosis and leishmaniosis among dogs in Ankara, Turkey. *Veterinary Parasitology*. 2005, v.129, p.187-191.

Avelino, Mm.; Campos Junior, D.; Parada, Jb.; Castro, Am. Risk factors for *Toxoplasma gondii* infection in women of childbearing age. *Braz. J. Infect. Dis*., 2004, v.8, n.2, Salvador, Apri. 2004.

Black, Mw. And Boothroyd, Jc. The lytic cycle of *Toxoplasma gondii. Microbiology and Molecular Biology Reviews,* 2000, v.64, p.607-623.

Brasil. Ministério da Saúde. Fundação Nacional de Saúde. *Boletim Epidemiológico,* v.2, n.3, 2002.

Bresciani, Kds.; Costa, Aj.; Toniollo, Gh.; Sabatini, Ga.; Moraes, Fr.; Paulillo, Ac.; Ferraudo, As. Experimental toxoplasmosis in pregnant bitches. *Veterinary Parasitology,* 1999, v.86, p.143-145.

Caporali, E.H.G.; Silva, A.V.; Mendonça, A.O.; Langoni, H. Comparação de métodos para determinação da prevalência de anticorpos anti-*Toxoplasma gondii* em suínos dos Estados de São Paulo e Pernambuco - Brasil. *Arquivos de Ciências Veterinárias e Zoologia da Unipar,* 2005, v.8, n.1, p.19-24.

Carruthers, VB. Armed and dangerous: *Toxoplasma gondii* uses and arsenal of secretory proteins to infect host cells. *Parasitology International*, 1999, v.48, p.1-10.

Chhabra, Mb.; Gupta, Sl.; Gautam, Op. Toxoplasma seroprevalence in animals in Northem Índia. *International journal of zoonoses*, 1985, v.12, p.136-142.

Chiari, Ca.; Neves, Dp.; Pereira Neves, D. Toxoplasmose humana adquirida através da ingestão de leite de cabra. *Memórias do Instituto Oswaldo Cruz*, 1984, v.79, n.3, p.337-340.

Clementino, Mm.; Souza, Mf.; Andrade Neto, Vf. Soroprevalence and *Toxoplasma gondii*-IgG avidity in sheep from Lajes, Brasil. *Veterinary Parasitology,* 2007, v.146, p.199-203.

Costa Aj.; Araújo, Fg.; Costa. Jo.; Lima, Jd.; Nascimento, E. Experimental infection of bovines with oocysts of *Toxoplasma gondii*. *Journal for Parasitology*, 1977, v.6, n.2, p.212-218.

Costa, Ghn.; Cabral, Dd.; Varandas, Np.; Sobral, Ea.; Borges, Fa.; Castagnolli, Kc. Freqüência de anticorpos anti-*Neospora caninum* e anti-*Toxoplasma gondii* em soros de bovinos pertencentes ao estados de São Paulo e Minas Gerais. Semina: *Ciências Agrárias*, 2001, v.22, n. 1, p. 57-62.

Da Silva, Av.; Culoto, Aa.; Langoni, H. Comparação da reação de Reação de Imunofluorescência indireta e do método de aglutinação direta na detecção de anticorpos anti-*Toxoplasma* em soro de ovino, caprino, canino e felino. *Arquivo do Instituto Biológico*, 2002, v. 69, n.1, p.7-11.

Daguer, H.; Vicente, Rt.; Costa, T.; Virmond, Mp.; Hamann, W.; Amendoeira, Mr. Soroprevalência de anticorpos anti-*Toxoplasma gondii* em bovinos e funcionários de matadouros da microrregião de Pato Branco, Paraná, Brasil. *Ciência Rural*, 2004, v.34, n.4, p.1133-1137.

Desmonts, G. e Remington, Js. Direct agglutination test for diagnosis of *Toxoplasma* infection: method for increasing sensitivity and specificity. *Journal of Clinical Microbiology*, 1980, v.11, p.562-568.

Derouin, F.; Lacroix, C.; Sumyuen, Mh.; Romand. S.; Garin, Yjf. Modelos experimentais de toxoplasmose: applications pharmacology. *Parasite*. 1995, v.2, p.243-256.

Desmonts, G. e Remington, Js. Direct agglutination test for diagnosis of *Toxoplasma* infection: method for increasing sensitivity and specRIFIcity. *Journal of Clinical Microbiology,* 1980, v.11, p.562-568.

Dubey, Jp. e Beattie, Cp. 1998. Toxoplasmosis of animals and man. Boca Raton*: CRC Press Inc.,* p.01-220., 1988.

Dubey, Jp. e Thulliez, P. Persistence of tissue cysts in edible tissues of cattle fed Toxplasma gondii oocysts. *American Journal Veterinary Research***,** 1993v.54, n.3, p.270-273.

Dubey, Jp.; Lindsay, Ds.; Speer, Ca. Structures of *Toxoplasma gondii* Tachyzoites, Bradyzoites, and Sporozoites and Biology and Development of Tissue Cysts. *Clinical Microbiology Reviews*, 1998, v.11, p. 67-69.

Dubey Jp, Miller Nl, Frenkel Jk. The *Toxoplasma gondii* oocyst from cat feces. *The Journal of Experimental Medicin***e**, 1970, v.132, p.636-62.

Dubey, .P.; Murrell, Kd.; Fayer, R.; Schab Ga. Distribution of *Toxoplasma gondii* tissue cysts in commercial cuts of pork. *J. Am. Vet. Med. Assoc.,* p.88, p.1035-1037, 1986.

Dubey, JP., Hill, DE., Jones, JL. et al. Prevalence of viable Toxoplasma gondii in beef, chicken and pork from retail meat stores in the United States; risk assessment to consumers. *J. Parasitol.*2005, v.91, p.1082–1093.

Dubey, Jp. Toxoplasmosis. *Journal of the American Veterinary Medical Association,* v.189, n.2, p.166-170, 1986.

Dubey, Jp. *Toxoplasma, Neospora, Sarcocystis* and other tissue cyst-forming of human and animals. In: KREIER, J.P. *Parasitic protozoa*. 2 ed., San Diego: Academic Press. p.1-157, 1993.

Dubey, Jp. Toxoplasmosis. *Journal of the American Veterinary Medical Association,* v.205, n.11, p.1593-1598, 1994.

Dubey, Jp. Pathogenicity and infectivity of *Toxoplasma gondii* oocysts for rats. *Journal of Parasitology*, v.82, p.951-956, 1996.

Dubey, Jp. Toxoplasmosis – a waterborne zoonosis. *Veterinary Parasitology*. v.126, p.57-72, 2004.

Dubey, Jp., 2007. *Toxoplasma gondii.* In: The Model Apicomplexan-Perspective And Methods, Weiss, L.D. and K. Kim (Eds.). Elsevier, London.

Faria, Eb.; Gennari, Sm.; Pena, Hfj.; Athayde, Acr.; Silva, Mlcr.; Azevedo, Ss. Prevalence of anti-*Toxoplasma gondii* and anti-*Neospora caninum* antibodies in goats slaughtered in the public slaughterhouse of Patos city, Paraíba State, Northeast region of Brazil. *Veterinary Parasitology*, 2007, v.149, p.126–129.

Frenkel, Jk.; Parker, Bb. An apparent role of dogs in the transmission of *Toxoplasma gondii.* The probable importance of Xenosmophilia. *Annals of the New York Academy of Sciences*, 1996, v. 791, p. 402-407.

Fulton Jd. and Turk, Jk. Direct aglutination test for *Toxoplasma gondii. Lancet*, 1959, v.2, p.1068.

Garcia, Jl.; Navarro, It.; Vidotto, O.; Gennari, Sm.; Machado, Rz.; Pereira, Abl.; Sinhorini, IL. *Toxoplasma gondii*: Comparison of a rhoptry-ELISA with IFAT and MAT for antibody detection in sera of experimentally infected pigs. *Experimental Parasitology*. 2006, v.113, n.2, June 2006, Pages 100-105.

Gauss, Cbl.; Dubey, Jp.; Vidal, D.; Ruiz, F.; Vicente, J.; Marco, I.; Lavin, S.; Gortazar, C.; Almería, S. Seroprevalence of *Toxoplasma gondii* in wild pigs (Sus scrofa) from Spain. *Veterinary Parasitology*, 2005, v.131, n.1-2, p.151-156.

Giessen, J.; Fonville, M.; Bouwknegt, M.; Langelaar, M.; Vollema, A. Seroprevalence of *Trichinella spiralis* and *Toxoplasma gondii* in pigs from different housing systems in The Netherlands. *Veterinary Parasitology*, 2007, v.148, p.371–374.

Giraldi, Jh.; Bracarense, Apfr.; Vidotto, O.; Tudury, Ea.; Navarro, It.; Batista, TN. Serology and histopathology of *Toxoplasma gondii and Neospora caninum* in dogs with neurologic disorders. *Semina: Ciências Agrárias*, Londrina, 2002, v.23, n.1, p. 9-14.

Goldsmith, Rs. 1998. Infectious Diseases: Protozoal and Helminthic In: *Current Medical Diagnosis and Treatment*. 37th Edition Stamford, Connecticut. USA: Appleton and Lange. 1998.

Gondim, Lfp., Barbosa Jr, H.V., Ribeiro Filho, Cha., Saeki, H. Serological survey of antibodies to *Toxoplasma gondii* in goats, sheep, cattle and water buffaloes in Bahia State, Brazil. *Veterinary Parasitology*, 1999, v.82, p.273-276.

Guimarães, Acs.; Coelho, Jr.; Kawarabayashi, M.; Requejo, Hiz.; Raymundo, Ml. Detecção de anticorpos IgG em pacientes com diferentes manifestações de toxoplasmose. *Revista Laes and Haes*, 1999, v.5, n.121.

Hall, S.; Ryan, M.; Buxton, D. 2001. The epidemiology of toxoplasma infection, In: Joynson, Hm.; Wreghitt, TG. *Toxoplasmosis. A Comprehensive Clinical Guide.* Cambridge University Press, p.58-124, 2001

Hiramoto, Rm. ; Mayrbaurl-Borges, M.; Galisteo Jr, Aj.; Meireles, Lr.; Macre, Ms; Andrade Jr, Hf. Infectivity of cysts of the ME-49 *Toxoplasma gondii* strain in bovine milk and homemade cheese. *Rev. Saúde Pública*, 2001, v.35, n.2, p.113-118.

Hutchison, Wm. Experimental transmission of *Toxoplasma gondii. Nature*, 1965, v.206, p.961-962.

James, GS. Comparison of cell culture, mouse inoculation and PCR for detection of *Toxoplasma gondii*: effects of storage conditions on sensitivity. *J. Clin. Microbiol.*, 1996, v.34, n.6, p. 1572-1575.

Janku, J. Pathogenesis and pathologic anatomy of coloboma of macula lutea in eye of normal dimensions, and in microphtalmic eye, with parasites in the retina. *Casopis Lekaru Ceskych,* 1923, v.62, p.1021-1027.

Jittapalapong, S. ; Sangvaranond, A.; Pinyopanuwat, N.; Chimnoi, W.; Khachaeram, S.; Koizumi, W.; Maruyama S. Seroprevalence of *Toxoplasma gondii* infection in domestic goats in Satun Province, Thailand. *Veterinary Parasitology*, 2005, v.127, n.1, p.23-28.

Jones, Jl.; Lopez, A.; Wilson, M.; Schulkin, J.; Biggs, R. Congenital Toxoplasmosis: A Review. *Obstetrical and Gynecological Survey*, 2001, v.56, n.5, p.296-305.

Langoni, H. et al. Inquérito soroepidemiológico para a toxoplasmose em ovinos no Estado de São Paulo, Brasil. *O Biológico*, São Paulo, 1999, v.61, n.1, p.35-39.

Lappin, Mr. Feline Zoonotic Diseases. *Veterinary Clinicals of North American Small Animal Practice,* 1993, v.23, n.1, p.57-77.

Lappin, Mr. *Doenças causadas por protozoários*. In: Ettinger, S. J.; Feldman, E. C. (Ed.). Tratado de medicina interna veterinária: doenças do cão e do gato. 5.ed. Rio de Janeiro: Guanabara Koogan, 2004. p.430-435.

Lindström, I; Kaddu-Mulindwa, D. H.; Kironde, F.; Lindh, J. Prevalence of latent and reactivated *Toxoplasma gondii* parasites in HIV-patients from Uganda. *Acta Trop.,* 2006, v.100, p.218-222.

Mead, Ps.; Slutsker, L.; Dietz, V.; Mccaig, L. F.; Bresee, Js.; Shapiro, C.; Griffin, Pm.; Tauxe, Rv. Food-related illness and death in the United States. *Emerging Infectious Diseases*, v.5, p.607-625, 1999.

Meireles, Lr. 2001. *Estudo das fontes de infecção da Toxoplasmose humana em diferentes localidades do Estado de São Paulo*. 2001. 171f. Dissertação (Mestrado em Ciências). – Instituto de Ciências Biomédicas da Universidade de São Paulo, São Paulo, 2001.

Miller, Nl.; Frenkel, Jk.; Dubey, Jp. Oral infections with Toxoplasma cysts and oocysts in felines, other mammals and birds. *Journal of Parasitology*, Lawrence, 1972, v.58, n.5, p.928-937.

Minho, Ap.; Freire, Rl.; Vidotto, O.; Gennari, Sm.; Manara, Em.; Garcia, Jl.; Navarro, It. Avaliação dos testes de Reação de Imunofluorescência indireta e aglutinação modificada para a detecção de anticorpos anti-*Toxoplasma gondii* em suínos infectados experimentalmente. *Pesquisa Veterinária Brasileira*, 2004, v.24, n.4, p.199-202.

Miró, G.; Montoya, A.; Jiménez, S.; Frisuelos, C.; Mateo, M.; Fuentes, I. Prevalence of antibodies to *Toxoplasma gondii* and intestinal parasites in stray, farm and household cats in Spain. *Veterinary Parasitology*, 2004, v.126, n.3, p.249-255.

Montoya, Jg.; Liesenfeld, O. *Toxoplasmosis.* Lancet, London, 2004, v.363, n.9425, p.1965-1976.

Munir, A.; Zaman, M.; Eltorky, M. *Toxoplasma gondii* pneumonia in a pancreas transplant patient. *South Med. Journal*, 2000, v.93, n.06, p.614-617.

Neto, Va.; Medeiros, Eas.; Levi, Gc.; Duarte, Mis. 1995. *Toxoplasmose,* São Paulo, Sarvier, 1995. 154p.

Nicolle, C.; Manceaux, L. Sur un protozoaire nouveau du *gondii. Comptes Rendus de l'Académie des Sciences,* 1909, v.148, p.369.

Pena Hfj.; Gennari, Sm.; Dubey, Jp.; Su, C. Population structure and mouse-virulence of *Toxoplasma gondii* in Brazil. *International Journal for Parasitology,* 2008, v.38, n.5, p.561-569.

Pereira-Bueno, J.; Quintanilla-Gonzalo, A.; Pérez-Perez, V.; Álvarez-Garcia, G.; Collantes-Fernández, E.; Ortega-Mora, Lm. Evaluation of ovine abortion associated with *Toxoplasma gondii* in Spain by different diagnostic techniques. *Veterinary Parasitology*. 2004, v.121, p.33-43.

Pinkerton, H. and Henderson, R.G. Adult toxoplasmosis. A previously unrecognized disease entry simulating the typhus-spotted ferver group. Journal American Medical Association, v.116, p.807-814, 1941. Apud Amato Net, V.; Medeiros, E.A.S.; Levi, G.C.; Duart, M.I.S. *Toxoplasmose.* 4.ed. São Paulo:, Savier, 1995, p.154.

Pinkerton, H.; Weinman, D. *Toxoplasma infection in man.* Archives of Pathology, Chicago, 1940, v.30, p.374.

Remington, Js.; Mcleod, R., Desmonts, G. 1995. *Toxoplasmosis.* In: Remington, J.S., Klein, J.O. eds. Infectious diseases of the fetus and newborn infant. 4ª ed. Philadelphia: WB Saunders, 140-167. 1995.

Roberts, F. and Mcleod, R. Pathogenesis of toxoplasmic retinochoroiditis. *Parasitol. Today,* 1999, v.15, n.2, p.51-57.

Roghmann, Mc.; Faulkner, Ct.; Lefkowitz, A.; Patton, S.; Zimmerman, J; Morris, Jg. Jr. Decreased seroprevalence for *Toxoplasma gondii* in

Seventh Day Adventists in Maryland. *American Journal of Tropical Medicine and Hygiene,* 1999, v.60, n.05, p.790-792.

Rorman, E.; Zamir, Cs.; Rilkis, I.; Ben-David, H. Congenital toxoplasmosis — prenatal aspects of *Toxoplasma gondii* infection. *Reproductive Toxicology*, 2006, v.21, n.4, p.458-472.

Sabin, Ab. and Feldman, Ha. Dyes as microchemical indicators of a new immunity phenomenon affecting a protozoon parasite (*T. gondii). Science*, Washington, 1948, v.108, p.660-663.

Santos Tr., Costa Aj., Toniollo Gh., Luvizotto Mcr., Benetti Ah., Santos Rr., Matta Dh., Lopes Wdz., Oliveira Ja., Oliveira Gp. Prevalence of anti-*Toxoplasma gondii* antibodies in dairy cattle, dogs, and humans from the Jauru micro-region, Mato Grosso state, Brazil. *Veterinary Parasitology*, v.161, p.324-326, 2009.

Santos Tr., Nunes, Cm., Luvizotto, Mcr., Moura, Ab., Lopes, Wdz., Costa, Aj. Bresciani, Kds. Detection of *Toxoplasma gondii* oocysts in environmental samples from public schools. *Veterinary Parasitology*, 2010, v.171, p. 53–57.

Sawadogo, P.; Hafid, J.; Bellete, B.; Sung, R. T. M.; Chakdi, M.; Flori, P.; Raberin, H.; Hamouni, I. B.; Chait, A.; Dalal, A. Seroprevalence of *T. gondii* in sheep from Marrakech, Morocco. *Veterinary Parasitology*, 2005, v.130, n.1-2, p.89-92.

Sikes, Rk. Toxoplasmosis. *Journal of the American Veterinary Medical Association*, 1982, v.180, n.8, p.57-859.

Silva, Dao.; Lobato, J.; Mineo, Twp.; Mineo, Jr. Evaluation of serological tests for the diagnosis of *Neospora caninum* infection in dogs: Optimization of cut off titers and inhibition studies of cross-reactivity with *Toxoplasma gondii. Veterinary Parasitology,* 2007, v.143, p. 234–244.

Souza, Aes.; Sousa, Dc.; Gomez, Jg.; Matos, Cs. Ocorrência de anticorpos anti-*Toxoplasma* em pacientes atendidos no Laboratório Celso Matos-Santarém, PA. *Revista Brasileira de Análises Clínicas*, 2002, v.34, n.1, p.51-52.

Souza, LM. 2001. *Neosporose e toxoplasmose bubalina: estudo da situação sorológica em rebanhos leiteiros do Estado de São Paulo*. 2001. 69f. Dissertação (Mestrado em Medicina Veterinária) – Faculdade de Ciência Agrárias e Veterinárias, Universidade Estadual Paulista, Jaboticabal, 2001.

Splendore, A. Un nuovo protozoa parasita dei Gonigli incontrato nelle lesioni anatomiche d'una malattia chericorda in multipunti il Kala-Azar dell

Uomo. *Revista da Sociedade Scientifica de São Paulo,* São Paulo, 1908, v.3, p.109-112.

Stalheim, Oh.; Hubbert, Wt.; Boothe, Ad.; Zimmermann, Wj.; Hughes, De.; Barnett, D.; Riley, J. L; Foley, J. Experimental toxoplasmosis in calves and pregnant cows. *Am. J. Vet. Res.,* 1980, v.41, p.10-13.

Tenter, Am.; Heckeroth, Ar.; Weiss, Lm. *Toxoplasma gondii*: from animals to humans. *International Journal for Parasitology*, 2000, v.30, n.12-13, p.1217-1258.

Torres, Cm. Affinité de I'Encephalitozoon chagasi agent étiologique d'une méningoencephalomyélitis congénitale avec myocardite et myosite chez I'homme. C. R. Soc. Biol., v.97, p.1797-1799, 1927. Apud: Amato Neto, V.; Medeiros, E. A. S.; Levi, G. C.; Duarte, M. I. S. *Toxoplasmose,* 4.ed. São Paulo, Savier, 1995, 154p.

Uchôa, C. M. A.; Duarte, R.; Laurentino-Silva, V.; Alexandre, G. M. C.; Ferreira, Hg. Padronização de ensaio imunoenzimático para pesquisa de anticorpos das calasses IgM e IgG anti-*Toxoplasma gondii* e comparação com a técnica de imunofluorescência indireta. *Revista da Sociedade Brasileira de Medicina Tropical*, 1999, v.32, p.661-669.

Vidotto, O.; Navarro, It.; Moco, Ca., et al. *Prevalência de Toxoplasma gondii em suínos abatidos em matadouros no norte do Paraná.* In: Encontro De Pesquisas Veterinárias, 2., 1986, Londrina.

Vidotto, O. Toxoplasmose: epidemiologia e importância da doença na saúde animal. *Semina: Ciências Agrárias*, 1992, v.13, p.69-75.

Wolf, A.; Cowen, D. Granulomatous encephalomyelitis due to an encephalitozoon (encephalitozic encephalomyelitis). A new protozoan disease of man. *Bulletin of the Neurological Institute of New York,* New York, 1937, v.6, p.306-371.

Wong, Sy. and Remington, JS. Biology of *Toxoplasma gondii. Aids*, 1993, v. 07, n.03, p.299-316.

Yu, J.; Xia, Z.; Liu, Q.; Liu, J.; Ding, J.; Zhang, W. Seroepidemiology of *Neospora caninum* and *Toxoplasma gondii* in cattle and water buffaloes (*Bubalus bubalis*) in the People's Republic of China. *Veterinary Parasitology*, , 2007, v.143, p.79-85.

In: *Toxoplasma Gondii* ISBN: 978-1-62618-349-0
Editors: M. Goodson and D. Winslow 

Chapter V

# *Toxoplasma Gondii*: The Role of Chickens in the Transmission of Toxoplasmosis to Humans

***Valter Ferreira de Andrade-Neto*** [*]
***and Ywlliane da Silva Rodrigues Meurer***
Laboratory of Malaria and Toxoplasmosis Biology,
Department of Microbiology and Parasitology, Center of Bioscience,
Universidade Federal do Rio Grande do Norte – UFRN,
Campus Universitário, Natal/RN, Brazil

## Abstract

Toxoplasmosis is an anthropozoonosis caused by the protozoan *Toxoplasma gondii* (*T. gondii*), an obligate intracellular parasite belonging to the phylum Apicomplexa, Sporozoea Class, Coccidia Sub-class, Eucoccidiida Order, Eimeriina Sub-order, Sarcocystidae Family and Toxoplasmatinae Sub-Family (Levine et al., 1988). It is a protozoan capable of infecting nucleated cells from a wide variety of mammals and

[*] Corresponding author: Laboratory of Malaria and Toxoplasmosis Biology, Department of Microbiology and Parasitology, Center of Bioscience, Universidade Federal do Rio Grande do Norte – UFRN, Campus Universitário, Av. Senador Salgado Filho, CEP 59078-970 Natal/RN, Brazil. Phone: +55 84 3215-3121, *E*-mail: aneto@cb.ufrn.br

birds as well as primary cell cultures of fish and insects (Mineo; Kasper, 1994). It presents a cosmopolitan geographic distribution, with high rates of infection in humans. It is estimated that one third of the world population has already been exposed to the *T. gondii* (Montoya, Liesenfeld, 2004). The first description of the species was carried out by Alfonso Splendore (1909 apud *Cox*, 2002) in Brazil in 1908, who reported the presence of the protozoan parasitizing laboratory rabbits. In that same year, the French parasitologists Charles Nicolle and Louis Hebert Manceaux (1908 apud *Cox*, 2002) isolated it in an African rodent (*Ctenodactylus gondii*), at the Pasteur Institute of Tunisia. Originally denominated *Leishmania gondii*, by its similarity with the protozoan of the genus *Leishmania*, the adequacy of its nomenclature only occurred in the following year. The life cycle of the *T. gondii* was only elucidated more than 50 years after the first report of its discovery (Evans, 1992). In the late 1960s, the analysis of cat fecal material made it possible to discover an evolutive type of protozoan, which was capable of inducing infection in other mammals and birds by its ingestion. Its coccidial cycle was elucidated in the 1970s, when the sexual stages of *T. gondii* were found in small intestine of cats (Frenkel et al. 1970; Dubey; Beattie, 1988). Although the protozoan *T. gondii* is considered the only valid species in its genus, three clonal genetic lineages were initially recognized. These are known as Type I, Type II and Type III, and they are predominant in Western Europe and the United States (Howe, Sibley, 1995; Howe et al., 1997). However, surveys on samples of this parasite from other regions of the world, such as South American localities, have revealed other genotypes, differing from the archetypes mentioned above (Soares et al., 2011; Clementino Andrade et al., 2013). When several markers were applied against Brazilian strains, a highly reticulated phylogenetic structure was identified, which suggests that recombination plays an important role in strain diversification in South America (Pena et al., 2008). Currently, it is known that its life cycle is divided between feline and non-feline infections, which are correlated with sexual and asexual replication, respectively (Black; Boothroyd, 2000). Definitive hosts are members of the family Felidae, for example domestic cats (Jackson; Hutchison, 1989; Dubey, 1986; 1993; Frenkel, 2000).

Asexual stages of toxoplasma-like parasites were first observed at the turn of the century in tissues of non-felines, for example birds and mammals (Dubey; Beattie, 1988). They are probably all warm-blooded animals including most livestock, and humans. In intermediate hosts, *T. gondii* undergoes two phases of asexual development. In the first phase, tachyzoites (or endozoites) multiply rapidly by repeated endodyogeny in many different types of host cells. Tachyzoites of the last generation initiate the second phase of development which results in the formation of tissue cysts. Within the tissue cyst, bradyzoites (or cystozoites) multiply slowly by endodyogeny. Tissue cysts have a high affinity for neural and muscular tissues (Ferguson, Hutchison; 1987). They are located predominantly in the central nervous system (CNS), the eye as well as skeletal and cardiac muscles. Unsporulated oocysts are released into the intestinal lumen and passed into the environment with the faeces. Sporogony occurs outside the host and leads to the development of infectious oocysts which contain two sporocysts, each containing four sporozoites (Dubey; Beattie, 1988; Frenkel, 2000). Several studies have been carried out on the asexual stages of *T. gondii*, in particular on the tachyzoite, while much fewer studies have considered the sexual stages or their infectious product, i.e. the sporozoites within the oocyst. It is noteworthy that cats and wild felids are essential to the persistence of *T. gondii* in hosts such as grazing animals (e.g. sheep, deer and, more recently, chickens) because they serve as the sole source of the infectious oocysts that contaminate the environment. Environmental contamination by *T. gondii* oocysts is a concern from both public health and biodiversity perspectives. Surface run-off containing feline feces containing *T. gondii* oocysts enters in multiple ecosystems (Dubey, 1986; Jackson, Hutchison; 1989). Currently ingesting undercooked meat also provides an important avenue for *T. gondii* infection through the ingestion of viable tissue cysts (Dubey et al, 1998a). Undercooked pork is another common risk factor for *T. gondii* exposure; beef is not believed to be a major player, however, mammals and avians raw meat was found to be a risk factor in several epidemiological studies in developed and underdeveloped countries (Fayer et al., 2004).

Even as other important risk factors in evidence are for infection include not washing kitchen knives after they have been used to cut raw meat, fruits, and vegetables, as well as infrequent hand washing (Jones et al, 2009). In general, the infection by the *T. gondii* has a higher prevalence in tropical or subtropical most climate zones, which favors the viability of oocysts in the environment (*Tenter* et al., 2000). The prevalence of reagents for anti-*T. gondii* antibodies increases with age and it varies depending on the cultural patterns of the population, its hygiene, food and urban and rural provenance (Amendoeira, 2003; Santos et al., 2012).

The prevalence of the disease in children is relatively low, increasing with age and the time of exposure to risk factors throughout life (Kompalic-Cristo et al., 2005). In Brazil, the official statistics from the Ministry of Health indicate toxoplasmosis as responsible for 90 deaths annually. However, there is a lack of data on morbidity and mortality caused by this parasite, so this data is formed by sub-estimated numbers.

*Toxoplasma Gondii* strain involved in the infection and host's immune status, the disease may cause severe illness in human beings and several other classes of animals (Dubey, Bettie, 1988; Cavalcante et al, 2006b). Over the last decades, toxoplasmosis has stimulated the interest of many researchers, mainly by public health relevance and by pervasive affected systems in the hosts, either as an opportunistic disease in immunocompromised patients and pregnant women, or as adaptative disease in immunocompetent animals where reproductive and neural systems is most affected (Tenter et al., 2000). Millions of people worldwide have a serological prevalence of between 15% and 85% in different regions of the planet (Kompalic-Cristo et al.; 2005). In the United States, the prevalence of seropositivity ranges from 10% to 30%, approximately 14% in the reproductive age. In Norway, for example, the positivity in serologic tests is only 10.9%. On the other hand, the prevalence is 28% in Denmark and it reaches 36.7% in Austria. In Malaysia, a recent study found a rate of prevalence of 49%, reaching rates of up to 70% in France, especially in Paris (Porto et al.; 2008). In Brazil, it is estimated that 70% of the population will manifest clinical signs of toxoplasmosis at some point in life. Surveys conducted in pregnant women have shown high seropositivity rates, ranging from 40% to 80% (Porto et al.; 2008).

The frequency of seropositive pregnant women for IgG anti-Toxoplasma varies widely among regions, from 28.1% to 70.9% in the South, 70.7% to 98.8% in the Midwest, 50.1% to 90.0% in the Southeast and 66.4% to 72.7% in the Northern Region of the country. In the Northeast, information about the frequency of infections in pregnant women, capable of perinatal transmission, is scarce; in the Sergipe State there was a prevalence rate of 69% in a study performed with 9.550 pregnant women, in Salvador city (BA) the rate was of 69.3%, 71.5% in Ceará, 74.7% in Recife (PE), in João Pessoa (PB) it was observed a rate of 91% of prevalence in a study performed with 55 pregnant women, 86.2% in Alagoas, 66.38% in São Luis (MA) and in Natal (RN) Barbosa et al (2009) observed a rate of prevalence of 66% in a study comprising 190 pregnant women. In the North of Country, in Belém (PA), a rate of 73% was obtained in a study conducted with 531 pregnant women, a rate of 76.2% in Amapá and reaching levels above 70% in Amazonas city. In the Southeast, in the capital city of the state of São Paulo the prevalence rate of 67.7% was observed in a study comprising 481 pregnant women and, in Rio de Janeiro, 77.1%; 73.5% in Espírito Santo and 51.6% in Uberlândia (MG).

In the South region, in the capital city of Rio Grande do Sul, a prevalence rate of 61.1% (Reis et al., 2006) was obtained by evaluating 10.468 pregnant women and the rate of 49.2% in Paraná. In the central region of the country, in Goiania, the prevalence rate has remained stable over the past 23 years. Nearly all species of animals are susceptible to infection by the protozoan, the seropositivity rates and clinical signs of toxoplasmosis in different hosts are varied (Dubey; Beattie, 1988).

# Chickens and Transmission of Toxoplasmosis to Humans

Chickens are considered one of the most important hosts in the epidemiology of *T. gondii* infection because they are an efficient source of infection for cats that excrete the environmentally resistant oocysts and because humans may become infected with this parasite after eating undercooked infected chicken meat (Dubey, 2010). Thereby, several species of domestic and wild birds have serological evidence for positive *T. gondii*, such as anseriformes (ducks), accipitriformes (hawks, buzzards, vultures and falcons), galiformes (partridges, pheasants, turkeys and chickens), gruiformes (coots), charadriiformes (gulls and terns), columbiformes (pigeons and doves), strigiformes (owls) and passerines (sparrows) (Dubey, 2002a,b). The infection of birds by *T. gondii* has been observed since 1911, when the parasites like *Toxoplasma* were observed by Carini (1911) on slides prepared from liver and spleen of a pigeon. Initially, these parasites found in birds were designated *T. columbae* (Yakimoff; Kohl-Yakimoff, 1912), *T. avium* (Marullaz, 1913), *T. fulicae*, according to the compromised host. Subsequently, Levine (1977) interchangeably appointed all *Toxoplasma* species which afflict poultry by *T. gondii*. Serological studies in poultry and isolation of *T. gondii* from tissues of birds by bioassays in mice and cats have been reported in several countries. These birds are from zoos (Literak et al., 1992; Zhang et al., 2000; Dubey et al., 2001; Dubey et al., 2009), private breeding sites (Dubey et al., 2004a; Leite et al., 2007) and free-living birds (Dubey et al. 1992; Quist et al. 1995; Work et al. 2000; Dubey et al. 2003a,b). Some bird species are resistant to infection by *T. gondii*, but others may develop symptoms of the disease. In the year 2000, two geese died of toxoplasmosis, presenting the parasite cysts in the liver, spleen and ureters, and significant injuries in other organs (Dubey et al., 2001). The tachyzoites of *T. gondii* were found, by immunohistochemistry, in the myocardium of an eagle, which died of necrotizing myocarditis (Szabo et al., 2004). Pigeons naturally infected with *T. gondii* showed clinical signs of anorexia, malaise and weight loss, culminating in seizures and death (Carini, 1911).

Another important aspect related to the prevalence of toxoplasmosis in birds is the role these animals play in the transmission of *T. gondii* to carnivores, behaving as reservoirs of the disease. Thus, domestic chickens (*Gallus gallus domesticus*) may play an important role in the epidemiology of toxoplasmosis. Several studies (Camargo et al., 1995; Cavalcante et al., 2006a;

Dubey et al., 2003a-b; Dubey et al., 2006) indicate the role of these birds as a direct source of infection for humans, especially those who maintain close contact to domestic productions, since this relationship may be linked to the increased opportunity of acquiring the infection through the consumption of parasitized meat and/or eggs. This species also assumes significant importance in the life cycle of *T. gondii*, since it is considered, along with other birds and small rodents, an important source of protozoan cysts for domestic cats due to the predatory habit of these animals (Dubey; Frenkel, 1998).

Several studies suggest an elegant strategy of transmission that is especially intriguing, a host manipulation hypothesis, which occurs when a parasite enhances its own transmission by altering host behavior (Banard, 1990). Transmission to the definitive host is mediated trophically, i.e., when a naive feline ingests an infected intermediate host through predation. Some trophically-transmitted parasites are capable of manipulating the behavior of their intermediate host in order to increase the chances of being predated by the definitive host and complete their life cycle (Cezilly, Thomas, et al., 2010; Adamo, 2013). Since sexual reproduction of *T. gondii* can be accomplished only in the definitive hosts, the parasite is ultimately dependent upon predation of infected prey by cats. As a result, there might be strong selective pressures for transmission to the definitive hosts and, thereby, the parasite evolved mechanisms to enhance the transmission rate from intermediate hosts to cats (Berdoy et al., 2000; Webster, 2001; Webster et al., 2006). A clear illustration of this manipulation hypothesis may be any equivalent increase in reaction times in a small rodent or bird intermediate host could have a significant impact upon their likelihood of predation by the feline definitive host, and hence would be subject to strong selective pressure under the original natural conditions (Webster, 2001). Likewise, even within potential intermediate hosts such as rodents, an example of a disadvantageous side-effect of parasite-altered behavior is that, although parasitised rodents may be more susceptible to predation, the predator may not be a suitable definitive host for the parasite (Berdoy et al., 1995). This has been suggested to be an unavoidable cost of altered behavior, as it could induce the emergence of different strains or clonal lineages (Moore, Gotelli, 1990).

The interest in detecting the prevalence of infection with *T. gondii* in domestic chickens is similarly connected to the environmental contamination. These animals are accustomed to eating directly from the ground and the result of this activity would be their infection by ingesting oocysts. The range chickens usually feed on scraps of food provided by the creators and also have habit of preying on small rodents, by which they can become infected by tissue cysts (Dubey, 2010). Together, these behavioral characteristics, combined with the high susceptibility of the host to the protozoan, these animals become significant epidemiological indicators of the agent in the environment (Devada et al., 1998; Dubey, 2010). Hence, domestic chickens are used as sentinel animals in regions of high prevalence of human infection (Da Silva et al., 2003). Studies concerning infections by *T. gondii* in chickens have received great attention in many regions of the world, including Brazil (Dubey, 2010). Direct detection of *T. gondii* oocysts in soil is difficult. Therefore, chickens have been used as one of the indicators of soil contamination by oocysts (Coutinho et al., 1982). Researchers initiated a worldwide survey of *T. gondii* infection in free-range chickens with the ultimate objective of studying the genetic diversity of *T. gondii* on a worldwide basis (Dubey et al., 2002; Lehmann et al., 2003, 2006; Dubey; Su, 2009).

Free-range chickens play an important role in the epidemiology of *T. gondii* in the rural environment, perhaps more than rodents, because they are clinically resistant to *T. gondii* and live longer than rodents. There are only a few reports of clinical toxoplasmosis in chickens worldwide. Chickens inoculated orally with *T. gondii* oocysts also remained asymptomatic or developed a mild illness (Miller et al., 1972; Biancifiori et al., 1986; Dubey et al., 1993; Kaneto et al., 1997). A recent review includes farming recommendations to reduce the prevalence of *T.gondii* infection of meat (Tenter et al, 2000). It is possible to markedly curtail infection of swine and poultry that are raised indoors and fed protected feedstuffs. Particular attention must be paid to preventing cats from having access to feedstuffs; contamination of any ingredient of a mixed ration with *T. gondii*-containing feces has the potential to rapidly disseminate infections within a herd. Attention should also be paid to ensure that water cannot be contaminated with feline feces. It is pertinent to note that the issue of contaminated meat may be of greater importance to animal industries than to individual farmers, because there is currently little economic incentive to ensure that animals sold to slaughter are free of this parasite (McAllister, 2005). The role of the industrial scale chicken breeding is of minor importance in relation to the toxoplasmic transmission to humans, due to the fast farming system, which does not allow the contact with felines.

However, it contrasts with the domestic small-scale creation, in which the poultry coexist in the same ecosystem for years (Araujo et al., 1989; Literak; Hejlicek, 1993). Although chickens bred in an intensive system apparently do not offer great risks to human contamination by ingesting tissue cysts of *T. gondii*, it is important to adopt the serological investigation in chickens raised in the "backyard", the increased consumption of chicken meat, associated to the cultural food practices, and the high susceptibility of the host to the infection by the protozoan, may provide reasons for the avian toxoplasmosis to exceed its relevance in the field of environmental pollution and veterinary healthcare, getting a substantial epidemiological importance in the transmission cycle of the disease (Cong et al., 2012; Dubey, 2010). Recent outbreaks of toxoplasmosis in humans in various regions of the world demonstrate that the sources of infection vary greatly in different human populations with differences in culture and eating habits (Dubey, 2010). It is likely that the major sources of *T. gondii* infections. However, it is also important to take into account that epidemiology is in a state of flux, in particular in the case of a versatile parasite, such as *T. gondii*. There are many factors that have an impact on the epidemiology of *T. gondii* infections, such as the type of management and production of livestock, hygienic standards of abattoirs, food processing and technology, the density of cats or wild felines in the environment, environmental conditions that have an influence on the sporulation of oocysts in the environment, (i.e. temperature, humidity, wind), geographical location (with respect to latitude) as well as the different habits of human consumers, to name just a few (Tenter, 2000; Dubey, 2010).

# References

Adamo, S. A. Parasites: evolution's neurobiologists. *J. Exp. Biol.* 216(Pt 1): 3-10. 2013.

Amendoeira, M.R.R.; Sobral, C.A.Q.; Teva, A.; Lima, J.N.; Klein, C.H. Inquérito sorológico para a infecção por *Toxoplasma gondii* em ameríndios isolados, Mato Grosso. *Rev. Soc. Bras. Med. Trop.,* vol. 36(6), 671-676, 2003.

Araujo F.A.P.; Silva N.R.S.; Chaplin E.L.; Bigatti, L. E._Prevalência de anticorpos toxoplásmicos em frangos abatidos para consumo humano em Porto Alegre, Rio Grande do Sul. *Arquivos da Faculdade de Veterinária da UFRGS*, v.17, p.23-28, 1989.

Barbosa, I.R., Holanda, C.M.C.X, Andrade-Neto, V.F. Toxoplasmosis screening and risk factors amongst pregnant females in Natal, northeastern Brazil. *Transactions of the Royal Society of Tropica Medicine and Hygiene*. Vol. 103.p.377-382. 2009.

Barnard, C.J. Parasitic relationships. C.J Barnard, J.M Behnke (Eds.), *Parasitism and Host Behaviour*, Taylor and Francis, London, pp. 1–33., 1990.

Berdoy, M., Webster, J.P., Macdonald, D.W. The manipulation of rat behavior by *Toxoplasma gondii. Mammalia.* v. 59. p. 605-614. 1995.

Berdoy, M., Webster, J.P., MacDonald, D.W., Fatal attraction in rats infected with Toxoplasma gondii. *Proceedings of the Royal Society of London B* 267, 1591-1594. 2000.

Biancifiori, F., C. Rondini, V. Grelloni, and T. Frescura,: Avian toxoplasmosis: experimental infection of chicken and pigeon. *Comp. Immunol. Microbiol. Infect. Dis.* 9, 337–346.1986.

Black, M. W.; Boothroyd, J.C. Lytic cycle of Toxoplasma gondii. *Microbiology and Molecular Biology Reviews*, v. 64, n. 3, p. 607-623, 2000.

Camargo, M. C. V.; Antunes, C. M. F.; Chiari, C. A. Epidemiologia da infecção por*Toxoplasma gondii* no município de Ribeirão das Neves, MG. I. Importância dos animais domésticos como fonte de infecção do T. gondii para o homem. *Revista da Sociedade Brasileira de Medicina Tropical*, v.28, n.3, p. 211-214, 1995.

Carini, A. Infection spontanée du pigeon et du chien due au Toxoplasma cuniculi. *Bull. Soc. Pathol. Exot.,* v. 4, p. 518–519, 1911.

Cavalcante, G.T., Aguiar, D. M., Camargo, L.M A., Labruna, M. B., Andrade, H. F. de, Meireles, L. R., Dubey, J. P., Thulliez, P., Dias, R. A., Gennari, S.M. Seroprevalence of Toxoplasma gondii Antibodies in Humans From Rural Western Amazon, Brazil. *Journal of Parasitology*: Vol. 92, No. 3, pp. 647-649. 2006a.

Cavalcante, G.T., Aguiar, D. M., Chiebao, D., Dubey, J.P., Ruiz, V.L.A., Dias, R.A., Camargo, L.M.A., Labruna, M.B., Gennari, S. M. Seroprevalence of Toxoplasma gondii Antibodies in Cats and Pigs From Rural Western Amazon, Brazil. *Journal of Parasitology.* Vol. 92, No. 4, pp. 863-864. 2006b.

Cezilly, F., Thomas, F. et al. Host-manipulation by parasites with complex life cycles: adaptive or not? *Trends Parasitol.* 26(6): 311-317. 2010.

Clementino Andrade, M.M., Pinheiro, B.V., Cunha, M.M., Carneiro, A.C.A.V., Andrade-Neto, V.F. New gentotypes of Toxoplasma gondii obtained from farm animals in Northeast Brazil. *Research in Veterinary Science.* 2013.

Cong, W., Huang, S.I., Zhou, D.H., Xu, M.J., Yan, C., Zhao, Q., Song, H.Q., Zhu, X.Q. Firt report of *Toxoplasma gondii* infection in market-sold adult chickes, ducks and pigeons in northwest China. *Parasites & Vectors*. v. 5 2012.

Coutinho S.G., Lobo R., Dutra G. Isolation of *Toxoplasma* from the soil during an outbreak of toxoplasmosis in a rural area in Brazil. *J Parasitol.* 1982.

Cox, F. E. G. History of human parasitology. *Clin. Microbiol. Rev.,* v. 15, n. 4, p. 595-612,2002.

Da Silva, D. S.; Bahia-Oliveira, L. M.; Shen, S. K., Kwok, O. C.; Lehman, T.; Dubey, J. P. Prevalence of *Toxoplasma gondii* in chickens from an area in southern Brazil highly endemic to humans. *The Journal of Parasitology,* v.89, p.394-396, 2003.

Devada, K.; Anandan, R.; Dubey, J. P. Serologic prevalence of *Toxoplasma gondii* in chickens in Madras, India. *Journal of Parasitology,* v.84, n.3, p.621-622, 1998.

Dubey, J. P. Toxoplasmosis in cats. Feline Pract, 16, pp. 12–45. 1986.

Dubey, J.P.; Beattie, C.P. *Toxoplasmosis of animals and man.* CRC Press, Boca Raton, FL. 1988.

Dubey, J.P.; Porter, S.L.; Tseng, F.; Shen, S.K.; Thulliez, P. Induced toxoplasmosis in owls. *Journal of Zoo and Wildlife Medicine*, v.23, p.98 - 102, 1992.

Dubey, J. P. Toxoplasma, Neospora, Sarcocystis, and other tissue cyst-forming coccidia of humans and animals, in: J. P Kreier (Ed.), Parasitic protozoa, *Parasitic protozoa* (2nd ed), pp. 1–158. 1993a.

Dubey, J. P., M. D. Ruff, M. E. Camargo, S. K. Shen, G. L. ilkins, O. C. H. Kwok, and P. Thulliez,: Serologic and parasitologic responses of domestic chickens after oral inoculation with Toxoplasma gondii oocysts. *Am. J. Vet. Res.* 54, 1668–1672. 1993b.

Dubey, J.P. Toxoplasmosis, sarcocystosis, isosporosis, and cyclosporosis. S.R Palmer, E.J.L Soulsby, D.J.H Simpson (Eds.), *Zoonoses*, Oxford University Press, Oxford, pp. 579–597. 1998a.

Dubey, J. P.; Frenkel, J. K. Toxoplasmosis of rats: a review, with considerations of their value as an animal model and their possible role in epidemiology. *Veterinary Parasitology*, v.77, n.1, p.1-32, 1998b.

Dubey, J.P.; Garner, M.W.; Willette, M.M.; Batey, K.L.; Gardiner, C.H. Disseminated toxoplasmosis in magpie geese (Anseranas semipalmata) with large numbers of tissue cysts in livers. *The Journal of Parasitology*, v.87, p.219-223, 2001.

Dubey, J. P. A review of toxoplasmosis in wild birds. Veterinary Parasitology, v.106,p.121-153, 2002a.

Dubey, J. P., Graham, D. H., Blackston,C. R., Lehmann, T.S., Gennari, M., Ragozo, A. M. A., Nishi, S.M.S., Shen, K., Kwok, O.C.H., Hill, D.E., Thulliez, P. Biological and genetic characterisation of *Toxoplasma gondii* isolates from chickens (Gallus domesticus) from São Paulo, Brazil: unexpected findings. *Int. J. Parasitol.* 32, 99–105. 2002b.

Dubey, J. P.; Navarro, I. T.; Graham, D. H.; Dahl, E.; Freire, R. L.; Prudencio, L. B.; Sreekumar, C.; Vianna, M. C.; Lehmann, T. Characterization of Toxoplasma gondii isolates from free-range chickens from Paraná, Brazil. *Veterinary Parasitology*, v.117, n.3, p.229-234, 2003a.

Dubey, J.P.; Graham, D.H.; Dahl, E.; Hilali, M.; El-Ghaysh, A.; Sreekumar, C.; Kwok, O.C.; Shen, S.K.; Lehmann, T. Isolation and molecular characterization of Toxoplasma gondii from chickens and ducks from Egypt. *Veterinary Parasitology*, v.114, p.89-95, 2003b.

Dubey, J. P.; Parnell, P. G.; Sreekumar, C; Vianna, M. C.; De Young, R, W.;Dahl, E.; Lehmann, T. Biologic and molecular characteristics of Toxoplasma gondiiisolates from striped skunk (Mephitis mephitis), Canada goose (Branta canadensis), blackwingedlory (Eos cyanogenia), and cats (Felis catus). *J. Parasitol.,* v. 90, n. 5, p. 1171-1174,2004.

Dubey, J. P.; Patitucci, A. N.; Su, C.; Sundar, N.; Kwok, O. C. H.; Shen, S. K. Characterization of Toxoplasma gondii isolates in free-range chickens from Chile, South America. *Veterinary Parasitology*, v.140, p.76-82, 2006.

Dubey, J. P.; Su, C. Population biology of Toxoplasma gondii: what's out and where did they come from. *Mem. Inst. Oswaldo Cruz* 104 (Suppl. I), 104, 190–195. 2009a.

Dubey, J. P.; Velmurugan, G.V.; Morales, J. A.; Arquedas, R.; Su, C. Isolation of Toxoplasma gondii from the keel-billed toucan (*Ramphastos sulfuratus*) from Costa Rica. *J. Parasitol.,* v. 95, n. 2, p. 467-468, 2009b.

Dubey, J. P. Toxoplasma gondii Infections in Chickens *(Gallus domesticus):* Prevalence, Clinical Disease, Diagnosis and Public Health Significance. *Zoonoses and Public Health*, v.57, n. 1, p. 60-73, 2010.

Evans, R. Life cycle and animal infection. D.O Ho-Yen, A.W.L Joss (Eds.), *Human toxoplasmosis*, Oxford University Press, Oxford, pp. 26–55. 1992.

Fayer, R., Dubey, J.P., Lindsay, D.S. Zoonotic protozoa: from land to sea. *Trends in Parasitology*. v.20. p. 531-36. 2004.

Ferguson DJ, Hutchison WM. An ultrastructural study of the early development and tissue cyst formation of Toxoplasma gondii in the brains of mice. *Parasitol. Res.*; 73(6):483-91.1987.

Frenkel, J.K. *Besnoitia wallacei* of cats and rodents: with a reclassification of other cyst-forming isosporoid coccidian. *J. Parasitol.,* 63, pp. 611–628. 1977.

Frenkel, J.K. Biology of *Toxoplasma gondii.* P Ambroise-Thomas, E Peterse (Eds.), Congenital toxoplasmosis: scientific background, clinical management and control, Springer-Verlag, Paris, pp. 9–25. 2000.

Fuentes, J.M. Rubio, C. Ramirez, J. Alvar. Genotypic characterization of *Toxoplasma gondii* strains associated with human toxoplasmosis in Spain: direct analysis from clinical samples. *Journal of Clinical Microbiology*, 39, pp. 1566–1570. 2001.

Howe, D.K., Sibley, C.D. *Toxoplasma gondii* comprises three clonal lineages, correlation of parasite genotype with human disease. *Journal Infectious Diseases*, 172:, pp. 1561–1566. 1995.

Howe, D.K., Honore, S., Derouin, F., Sibley, L.D. Determination of genotypes of Toxoplasma gondii strains isolated from patients with toxoplasmosis. *Journal of Clinical Microbiology*, 35 pp. 1411–1414. 1997.

Jackson, M.H, Hutchison, W.M. The prevalence and source of *Toxoplasma* infection in the environment. *Adv. Parasitol.,* 28, pp. 55–105. 1989.

Jones, J.L., Dargelas, V., Roberts, J., Press, C., Remington, J.S., Montoya, J.G. Risk Factors for Toxoplasma gondii infection in the United States. *Clin. Infect. Dis.* 49, 878–884, 2009.

Kaneto, C. N., A. J. Costa, A. C. Paulillo, F. R. Moraes, T. O. Murakami, and M. V. Meireles,: Experimental toxoplasmosis in broiler chicks. *Vet. Parasitol.* 69, 203–210. 1997.

Kompalic-Cristo, A., Britto, C., Fernandes, O. Diagnóstico molecular da toxoplasmose: Revisão. *J Bras Patol Med Lab*, vol. 41 p. 229-35, 2005.

Lehmann, T., Graham, D. H. , Dahl, E., Sreekumar, C., Launer, F., Corn, J. L., Gamble, H.R., Dubey, J.P. Transmission dynamics of Toxoplasma gondii on a pig farm. *Infect.Genet. Evol.* 3, 135–141. 2003:

Lehmann, T., Marcet, P. L., Graham, D. H., Dahl, E. R., Dubey, J.P. Globalization and the population structure of Toxoplasma gondii. *Proc. Natl. Acad. Sci. U S A* 103, 11423–11428, 2006.

Leite, A.S.; Alves, L.C.; Faustino, M.A.G. Serological survey of Toxoplasmosis in Birds from Cracidae family in a wild bird center facility at Pernambuco state, Northeast of Brazil. *Medicina Veterinária Recife*, v.1, p.55-57, 2007.

Levine, N.D. Taxonomy of *Toxoplasma. J. Protozool.,* v. 24, p. 36–41, 1977.

Levine, N.D. The protozoan phylum apicomplexa: Vol. IICRC Press, Boca Raton, FL. 1988.

Liesenfeld, O. Oral infection of C57BL/6 mice with Toxoplasma gondii: a new model of inflammatory bowel disease? *J. Infect. Dis.,* v. 185, n. 1, p. 96-101, 2002.

Literák, I.; Hejlícek, K.; Nezval, J.; Folk, C. Incidence of *Toxoplasma gondii* in populations of wild birds in the Czech Republic. *Avian Pathology*, v.21, p.659- 665, 1992.

Marullaz, M. Au sujet d'un toxoplasme des oiseaux. *Bull. Soc. Pathol. Exot.*, v. 6, p.323–326, 1913.

McAllister, M. M. A decade of discoveries in veterinary protozoology changes our concept of ''subclinical'' toxoplasmosis. *Veterinary Parasitology*. 2005.

Miller, N. L., Frenkel, J. K., Dubey, J. P. Oral infections with Toxoplasma cysts and oocysts in felines, other mammals, and in birds. *J. Parasitol.* 58, 928–937. 1972.

Mineo, J. R.; Kasper, L. H. Attachment of *Toxoplasma gondii* to host cells involves major surface protein, SAG-1 (P30). *Exp. Parasitol.*, v. 79, p. 11-20, 1994.

Montoya, J.G., Liesenfeld, O. Toxoplasmosis. *The Lancet.* v.363, p.1965-1976. 2004.

Moore, J., Gotteli, N.J. Phylogenetic perspective on the evolution of altered host behaviours: a critical look at the manipulation hypothesis. In: C.J Barnard, J.M Behnke (Eds.), *Parasitism and Host Behaviour*, Taylor and Francis, London, pp. 193–230, 1990.

Pena, H.F.J., Gennari, S.M., Dubey, J.P., Su, C. Population structure and mouse-virulence of *Toxoplasma gondii* in Brazil. *International Journal for Parasitology*, 38, pp. 561–569. 2008.

Porto, A.M.F.; Amorim, M.M.R.; Coelho, I.C.N.; Santos, L.C. Perfil sorológico para toxoplasmose em gestantes atendidas em maternidade. *Rev. Asso. Med. Bras*. São Paulo, vol. 54, n. 3, 2008.

Quist, C.F.; Dubey, J.P.; Luttrell, M.P.; Davidson, W.R. Toxoplasmosis in wild turkeys: a case report and serologic survey. *Journal of Wildlife Diseases,* v.31, p.255-258, 1995.

Reis MM, Tessaro MM, d'Azevedo PA. Perfil sorológico para toxoplasmose em gestantes de um hospital público de Porto Alegre. *Rev Bras Ginecol Obstetr.*28:158-64; 2006.

Santos, L.S.S.; Carvalho, A.M.; Santos, L.M.J.F.; Cademartori, B.G.; Farias, N.A.R. Avaliação soroepidemiológica de Toxoplasma gondii em humanos que convivem com gatos (Felix catus) como animal de estimação.

Disponível em: http://www.ufpel.edu.br/enpos/2011/anais/pdf/CB/CB_00213.pdf Acesso em março 2013.

Sibley, L.D., Boothroyd, J.C. Virulent strains of *Toxoplasma gondii* comprise a single clonal lineage. Nature (London), 359, pp. 82–85. 1992.

Soares, R.M., Silveira, L.H, Silva, A.V., Ragozo, A., Galli, S., Lopes, E.G., Gennari, S.M., Pena, H.F.J. Genotyping of *Toxoplasma gondii* isolates from free range chickens in the Pantanal area of Brazil. *Veterinary Parasitology*. v. 178. p.29-34. 2011.

Su, C., Zhang, X., Dubey, J.P. Genotyping of Toxoplasma gondii *by* multilocus PCR-RFLP markers: a high resolution and simple method for identification of parasites. *International Journal for Parasitology*, 36, pp. 841–848. 2006.

Suzuki, Y.; Wong, S.Y.; Grumet, F.C.; Fessel, J.; Montoya, J.G.; Zolopa, A.R.; Portmore, A.; Schumacher-Perdreau, F.; Schrappe, M.; Koppen, S. Evidence for genetic regulation of susceptibility to toxoplasmic encephalitis in AIDS patients. *The Journal Infectious Diseases*, 173, pp. 265–268. 1996.

Szabo, K. A.; Mense, M. G.; Lipscomb, T. P.; Felix, K. J.; Dubey, J. P. Fatal toxoplasmosis in a bald eagle (Haliaeetus leucocephalus). *Journal of Parasitology*, v. 90, n.4, p. 907-908, 2004.

Tenter, A.M., Heckeroth, A.R., Weiss, L.M. T*oxoplasma gondii*: from animals to humans. *International Journal for Parasitology*, 30, pp. 1217–1258. 2000.

Webster, J.P. Rats, cats, people and parasites: the impact of latent toxoplasmosis on behaviour. *Microbes and Infection* 3, 1037-1045. 2001.

Webster, J.P., Lamberton, P.H.L., Donnelly, C.A., Torrey, E.F. Parasites as causative agents of human affective disorders? The impact of anti-psychotic, mood-stabilizer and anti-parasite medication on *Toxoplasma gondii*'s ability to alter host behaviour. *Proceedings of the Royal Society of London* B 273, 1023-1030. 2006.

Work, T. M.; Massey, J. G.; Rideout, B. A.; Gardiner, C. H.; Ledig, D. B.; Kwok, O. C. H.; Dubey, J. P. Fatal toxoplasmosis in free-ranging endangered 'Alala from Hawaii. *Journal of Wildlife Diseases*, v. 36, p. 205–212, 2000.

Yakimoff, W.L.; Kohl-Yakimoff, N. Toxoplasma canis (Mello). *Arch. Protistenk.*, v. 27, p. 195–206, 1912.

Zhang, S.Y.; Wei, M.X.; Zhou, Z.Y.; Yu, J.Y.; Shi, X.Q. Prevalence of antibodies to *Toxoplasma gondii* in the sera of rare wildlife in the

Shanghai Zoological Garden, People's Republic of China. *Parasitology International*, v.49, p.171-174, 2000.

In: *Toxoplasma Gondii*
Editors: M. Goodson and D. Winslow
ISBN: 978-1-62618-349-0


*Chapter VI*

# Epidemiology and Control of Toxoplasmosis in Cats

***Katia Denise Saraiva Bresciani[1,2*], André Luiz Baptista Galvão[2], Amanda Leal de Vasconcellos[2], Jancarlo Ferreira Gomes[3], Thaís Rabelo dos Santos[1,2], Rodrigo Rabelo dos Santos[2], Milena Araúz Viol[1], Monally Conceição Costa de Aquino[1] and Italmar Teodorico Navarro[4]***

[1]Faculdade de Medicina Veterinária de Araçatuba, UNESP, Araçatuba, São Paulo, Brazil
[2]Faculdade de Ciências Agrárias e Veterinárias, UNESP, , Brazil
[3]Instituto de Biologia e Instituto de Computação da Universidade de Campinas, SP, Brazil
[4]Universidade Estadual de Londrina, Paraná, Brazil

## Abstract

Toxoplasmosis, a zoonosis of worldwide distribution, has importance in human and veterinary medicine. Animals can be direct or indirect

[*] Email: bresciani@fmva.unesp.br

source of infection to man. The infection can be acquired by ingestion of tissue cysts in infected meat, ingestion of food or water contaminated with oocysts and congenitally, by tachyzoites crossing the placenta from the infected mother to the fetus. Cats may act as intermediate or definitive hosts of this parasite. Among the infected animals, felines have an important role in the epidemiology of this agent because like definitive hosts eliminate oocysts in the environment. Evaluations of the occurrence of *Toxoplasma gondii* antibodies in felines has drawn the attention of research worldwide. The aim of this chapter wil be clear some clinical manifestations aspects, address the true participation of cats in the spread of the disease, as well as discuss methods of diagnosis, therapeutic measures, prophylaxis and control of this disease.

## Introduction

The felines, definitive hosts of *Toxoplasma gondii*, can eliminate fecal oocysts, evolutionary forms improved by natural selection to support environmental adversities and can be highly infectious for several species of hosts.

In small animals, the clinical manifestations of toxoplasmosis are varied and similar to other diseases, with involvement of multiple organs and systems, such as the gastrointestinal, lymphatic, hematopoietic, hepatic, respiratory, muscular-skeletal, cardiovascular, ocular and nervous (Lappin et al., 2004). Cats defecate on soft earth or sand and bury their feces, which are consistent and can remain in the place for months. Unless the cat is ill, little or no faecal residue is adhered to the perianal region and in general, do not present diarrhea during the period of oocysts excretion. Because of its careful cleaning habits, fecal matter is not found in the pelage of clinically normal animals (therefore the possibility of transmission to humans by the act of touching or petting a cat is minimal or nonexistent).

It is important to pay attention to these peculiarities of feline behavior and also emphasize that the oocysts are eliminated by these definitive hosts, but need to stay in the environment for at least one day, under certain conditions of temperature and humidity and sporulate to become infective.

These informations are essential to guide the community, in general, about the real role that cats take in the cycle of this important zoonosis.

# Feline Toxoplasmosis

The elucidation of the life cycle of *T. gondii*, with the finding that elimination, by cats, of infective evolutionary forms, was possible through the studies of Hutchison, Dunachie and Work (1968), Hutchison et al. (1970, 1971), Piekarski and Witte (1970) and Frenkel (1973). In the context of parasite biology, the cat may act as definitive host (Frenkel et al., 1970) or intermediate and shelter extraintestinal stages (Dubey et al., 1996; Urquhart et al., 1998, Dubey, 2002).

The ingestion of tissue cysts, oocysts from infected cat feces and the congenital infection are the three main forms of *T. gondii* transmission. Other less important forms include ingestion of contaminated milk, transfer of body fluids and organ transplantation (Dubey, 1986; Dubey et al., 1990; Powell et al., 2001; Miró et al., 2004).

Dubey et al. (2004), isolated 37 *T. gondii* samples from 54 domestic cats from Paraná, in the city of Santa Isabel do Ivaí, during an outbreak of human toxoplasmosis. These isolates were 15 from type I and 22 from type III. Most isolates were virulent in mice, with no difference in relation to the genotype.

In São Paulo, 47 samples of *T. gondii* from 71 cats, 34 (72.4%) were type I, 12 (25.5%) were type III and 1 (2.1%) had mixed infection with genotypes I and III. In bioassay in mice, there was 97% mortality in animals infected with genotype type I and 42% with type III (Pena, 2004), with significant differences between them.

Immunosuppression has been observed in cats with toxoplasmosis and concomitant infections by feline leukemia virus (FeLV) and feline immunodeficiency virus (FIV). Particularly this last agent, suppresses CD4 + and CD8 +, reduces the expression of interleukins (IL-2, IL-6, TNF-γ), reactivate latent disease and causes re-shedding of fecal oocysts by cats (Lucas et al., 1998; Dorny et al., 2002; Dubey; Lappin, 2006; Swinger et al., 2009; Akhtardanesh et al., 2010).

In cats, in the northwestern part of São Paulo State, Bresciani et al. (2007), observed a higher incidence of *T. gondii* antibodies with aging. Coelho et al. (2011) demonstrated correlation of the disease with postnatal infection, type of diet and living with rodents. Sobrinho et al. (2012) found no association of this protozoan with visceral leishmaniasis in this animal species.

Results of studies about the frequency of *T. gondii* infection in cats in different regions of the world over the past 15 years are summarized in Table 1.

**Table 1. Occurrence of *Toxoplasma gondii* infection in cats, in the period from 1999 to 2004, in different countries**

| Locality | Seropre-valence (%) | Method* | Reference |
|---|---|---|---|
| Brazil | 73.0 | IFAT | Garcia et al. (1999) |
| Brazil | 46.0<br>43.4 | IFAT - IgG<br>IFAT - IgM | Carvalho (1999) |
| Brazil | 19.4 | IFAT | Langoni et al. (2001) |
| Brazil | 26.3 | MAT | Silva et al. (2002) |
| Brazil | 84.4 | MAT | Dubey et. al. (2004) |
| Brazil | 35.4 | MAT | Pena (2004) |
| Brazil | 17.0<br>40.0 | Indirect Hemagglutination<br>ELISA | Meireles et al. (2004) |
| Australia | 39.0 | ELISA | Summer, Ackland (1999) |
| United States | 42.0 | MAT | De Feo et al (2002) |
| United States | 48.0 | MAT | Dubey et al (2002b) |
| Belgium | 70.2 | Aglutinação Direta | Dorny et al. (2002) |
| Spain | 45.0 | MAT | Gauss et al. (2003) |
| Spain | 32.3 | IFAT | Miró et al. (2004) |
| Brazil | 25.0 | IFAT | Bresciani et al. (2007) |
| Brazil | 16.3 | IFAT | Cruz et al. (2011) |
| Brazil | 50.5 | IFAT | Braga et al. (2012) |

*IFAT: Indirect Fluorescent Antibody Test; ELISA: Enzyme Linked Immunosorbent Assay; MAT: Modified Agglutination Test.

# Clinic and Pathogenesis

Experimental infections in cats are often asymptomatic, few animals get sick and deaths rarely occur (Omata et al., 1990; Sato et al., 1993).

In 100 cats diagnosed with toxoplasmosis, confirmed by histopathology, were observed lung infections, neuropathy, hepatitis, pancreatitis, cardiovascular and ophthalmic changes in 97.7%, 96.4%, 93.3%, 84.4%, 86.4 % and 81.5% of cases, respectively (Dubey, Carpenter, 1993).

Common clinical manifestations in toxoplasmosis are fever and uveitis (Lappin, 2002; Lappin et al., 2004). The last one can be induced by cytolysis, via intraocular tachyzoites replication or by deposition of immune complexes.

In the ocular form, the agent invades the uveal tract, retina and optic nerve triggering inflammation with exudative retinal displacement (Lappin et al., 2004). Iridocyclitis, iritis, chorioretinitis, aqueous flare, keratic precipitates, crystalline lens dislocation, retinal detachment and glaucoma may occur concurrently or separately in cats without systemic involvement (DUBEY; Lappin, 2006).

Kittens infected via placenta or lactogenic ways may show inflammatory processes that affect the liver, lungs and central nervous system, which reflects clinically as uveitis, lethargy, depression, ascites, encephalitis, hypothermia and sudden death. In adults with primary or recurrent infection, are described anorexia, lethargy, dyspnea, pneumonia, intermittent fever, emaciation, vomiting, diarrhea, hyperesthesia, stiff gait, lameness, neurological deficits, dermatitis and death (Dubey; Lappin, 2006).

Tissue samples of cutaneous nodules on the body of a female feline, were positive for *T. gondii* by the reaction of immunohistochemistry and polymerase chain reaction (PCR), according to Anfray et al. (2005). Park et al. (2007) described the case of a cat that presented, in a mammary gland, a nodule of about three centimeters in diameter. Histopathologic examination showed the same pyogranulomatous necrotizing panniculitis, vasculitis and mastitis with clusters of tachyzoites, suggestion of infection by the parasite and the confirmation of the clinical suspicion was performed by molecular techniques.

Abortion and neonatal mortality were reported in pregnant cats inoculated with infective evolving *T. gondii* forms (Dubey, Hoover, 1977; Dubey, 1977; Sato et al., 1993; Dubey et al., 1996; Powell et al., 2001; Sakamoto et al., 2009).

Jaundice, cholangitis, ascites, hepatitis and pancreatitis may be associated with hepatobiliary compromise, promoted by *T. gondii*, which can promote accumulation of periportal mononuclear cells (Smart et al., 1973; Harvey; Greeffydd-Jones, 2010).

Dubey and Powell (2012) reported the occurrence of six million *T. gondii* oocysts in a fecal sample from a cat, male, six years old, with lethargy, anorexia, fever and diarrhea and positive serology for *T. gondii* of 1:800 by the modified agglutination test. Genotyping evidenced the type I, being the first isolation of this parasite genetically characterized, in a cat with acquired toxoplasmosis in the United States of America (USA).

In acute toxoplasmosis, can be noticed nonregenerative anemia, leukocytosis with neutrophilia or lymphocytosis, monocytosis and eosinophilia (Dubey; Lappin, 2006). Under conditions of pancreatitis, there is elevation of serum amylase and lipase as well as reduction in serum calcium (Dubey; Lappin, 2006).

In situations of tissue impairment, there may be hypoproteinemia, hypoglobulinemia, increased serum enzymatic activities of alanine amino transferase (ALT), alkaline phosphatase (ALP), bilirubin (Corrêa; Corrêa, 1992), with hypergammaglobulinemia in chronic phase (Dubey; Lappin, 2006).

Once infected, the animals may shelter bradyzoites in the tissues, which encourages a long-term humoral immune response. The detection of antibodies against *T. gondii* by indirect immunofluorescence as well as by direct agglutination test, hemagglutination test and enzyme immunoassay, are recommended methods for diagnosis and epidemiological survey of this zoonosis (Zhang et al., 2001; Dubey; Lappin, 2006).

In parasitological diagnosis by the Sheather method (centrifugal-flotation in sucrose) can be identified oocysts in the feces of cats. These evolutionary forms of coccidia isosporianos are morphologically indistinguishable on microscopic examination. Thus *T. gondii*, *Sarcocystis*, *Frenkelia* and *Besnoitia* can only be differentiated by molecular techniques (Dubey; Lappin, 2006).

The occurrence of fecal *T. gondii* oocysts is low because these are released only for one to two weeks during the clinical evolution of the disease, consisting of an occasional finding. Asymptomatic cats or suffering from diarrhea can excrete them (Dubey; Lappin, 2006).

*T. gondii t*achyzoites can be detected by cytology of body tissues and fluids samples, such as blood, cerebrospinal fluid, peritoneum and bronchoalveolar lavage samples. The tachyzoites may also be isolated from samples of ascitic fluid or pleural effusion (Brownlee; Sellon, 2001; Dubey; Lappin, 2006).

In acute progression of the disease, the radiographic examination of the thorax, it is expected to find diffuse interstitial *alveolar* pattern with lobar distribution. In severely compromised animals, it is possible to notice a diffuse and homogeneous pattern due to alveolar coalescence. In abdominal radiography are described intestinal masses, increased mesenteric lymph nodes and ascites; in ultrasonography there is tissue or organ enlargement, suggestive of granuloma formation. Myelography, computed tomography and magnetic resonance imaging can detect compromise of the central nervous system (Dubey; Lappin, 2006).

At necropsy, gross lesions are cerebral necrosis, discoloration of nervous tissue, cerebellar necrosis and atrophy, pulmonary necrosis, hepatic and mesenteric lymph nodes necrosis and the most common histopathological findings are encephalitis, nonsuppurative menigoencephalomyelitis, necrosis, malacia and gliosis, with eventualcompromise of peripheral nerves (Dubey; Lappin, 1998). However, there are no clinicopathological changes or pathognomonic diagnostic imaging for this disease and confirmation of the diagnosis depends on the isolation of the parasite, organism histopathologic demonstration in the lesions and positive serodiagnosis (Lappin et al., 2004 and Carvalho, 2007).

The clindamycin hydrochloride has good intestinal absorption, leading to clinical improvement within 24 to 48 hours after the beginning of its administration, with return of appetite, disappearance of fever and hyperesthesia and, with the continuity of administration, after weeks, there are reports of restablishment of the operation of cranial and spinal nerves, when compromised, as well as the resolution of polymyositis. In cases with ocular changes, clinical improvement is described after a week of the product use. In excessive doses, may cause local irritation in the digestive system, with side effects such as anorexia, vomiting and diarrhea (Dubey; Lappin, 2006).

This drug, orally or parenterally, is considered effective to treat the polysystemic disease and can shorten the elimination of oocysts by cats, acts on the blood brain barrier, the vascularity of the organs and tissues which confers patient clinical improvement, being used for control of toxoplasmosis in animals and humans, indicated for the treatment of pregnant females and can shorten the period of oocysts elimination for cats (Lappin et al., 2004; Swinger et al., 2009).

The use of glucocorticoids is contraindicated, since it can exacerbate the manifestation of the disease when there is systemic compromising, however, in case of injury in the organ of sight, the topical use of the drug, associated with clindamycin hydrochloride gives good result (Dubey; Lappin, 2006; Harvey; Greeffydd-Jones, 2010). When there is bone marrow compromising, folic acid supplementation is recommended (Dubey; Lappin, 2006).

The combination of trimethoprim-sulfametaxol has proven satisfactory effects on the control of toxoplasmosis, especially in the treatment of ocular toxoplasmosis in human (Dubey; Lappin, 2006). Papich (2009) described that the combination sulfadiazine-trimethoprim and trimethoprim-sulfametaxol combination are effective in the treatment and control of toxoplasmosis in small animals.

Barbosa et al. (2012) conducted studies *in vitro* and *in vivo* in experimental models using enrofloxacin in the treatment of toxoplasmosis, which proved effective with few side effects, may being an important alternative in the treatment of clinical toxoplasmosis manifestation.

Daily cleaning and the adequate removal of feces should be performed, once the oocysts require at least 24 hours to sporulate and become infectious.. It is recommended to avoid the habit of coprophagia observed in some animals, as well as the control of insects and rodents in the environment. Do not consume and do not provide to the animals raw or undercooked meat or raw milk from cow, sheep and goat, being indicated the use of commercial products or foods available. Personal hygiene behavior must be strictly followed and water from lakes, rivers and wells that do not undergo a pretreatment should not be consumed or provided to animals (Monteiro, 2010).

# Conclusion

Toxoplasmosis it is a zoonosis of worldwide distribution that promotes varied clinical symptoms, without clinicopathological findings or specific pathognomonic, thus forming, a clinical challenge to the veterinary in feline clinic. Although is possible that the cycle of *T. gondii* perpetuate in the absence of cats, it seems that this species is of essencial importance in the transmission of toxoplasmosis in different locations worldwide. Thus, the veterinarian plays an important role in guiding owners of small pets about relevant concepts in the control of this disease.

# References

Anfray, P.; Bonetti, C.; Fabbrini, F.; Magnino, S.; Manciani, F.; Abramo, F. Feline cutâneos toxoplasmosis a case report. *Veterinary Dermatology*, Oxford, v. 16, n. 2, p. 131-136, 2005

Akhtardanesh, B.; Ziaali, N.; Sharifi, H.; Rezaei, S. Feline immnodeficiency vírus, feline leukemia vírus and *Toxoplasma gondii* stray and household cats in Keman-Iran: Seroprevalence and correlation with clinical and laboratory findings. *Research in Veterinary Sciense*, London, v. 89, n. 2, 306-310, 2010.

Barbosa, B. F.; Gomes, A. O.; Ferro, E. A.; Napolitano, D. R.; Mineo, J. R.; Silva, N. M. Enrofloxacin in able to control *Toxoplasma gondii* infection in both in vitro and in vivo experimental models. *Veterinary Parasitology*, Amesterdam, v. 187, n. 1-2, p. 44-52, 2012.

Braga, M . S. C.; André, M. R.; Jusi, M. M. G.; Freschi, C. R.; Teixeira, M. C. A.; Machado, R. Z. Ocurrence of anti-*Toxoplasma gondii* and anti-Neospora caninum antibodies in cats with outdoor acess in São Luís, Maranhão, Brazil. *Revista Brasileira de Parasitologia*, Jaboticabal, v. 21, n. 2, p. 107-111, 2012.

Bresciani, K. D. S.; Costa, A. J.; Nunes, C. M.; Serrano, A. C. M.; Moura, A. B.; Stobbe, N. S.; Perri, S. H. V.; Dias, R. A.; Gennari, S. M. Ocorrência de anticorpos contra *Neospora caninum* e *Toxoplasma gondii* e estudo de fatores de risco em cães de Araçatuba (SP). *ARS Veterinária,* Jaboticabal, v. 23, n. 1, 2007.

Bresciani, K. D.S.; Gennari, S. M.; Serrano, A. C. M.; Rodrigues, A. A. R.; Ueno, T.; Franco, L. G.; Perri, S. H. V.; Amarante. Antibodies to Neosporoa caninum and *Toxoplasma gondii* in domestic cats from Braszil. *Parasitology Research*, New York, v. 100, n. p. 281-285, 2007.

Brownlee, L.; Sellon, R. K. Diagnosis of naturally ocorring toxoplasmosis by bronchoalveolar lavage in cat. *Journal of the American Hospital Association*, Schaumburg, v. 37, n. 3, p. 251-255, 2001.

Carvalho, C.S. Padronização do Elisa Teste para a detecção de anticorpos das classes IgM e IgG em soros de gatos experimentalmente infectados com taquizoítos de *T. gondii*. Dissertação (Mestrado em Medicina Veterinária) Universidade Estadual Paulista – Faculdade de Ciências Agrárias e Veterinárias. 76 p. 1999.

Carvalho, A. C. F. *Toxoplasmose: Morfologia e Morfometria da medula espinhal de cães soropositivos assintomáticos*. 2007. Tese de Doutorado (Doutorado em Medicina Veterinária) – Faculdade de Ciências Agrárias e Veterinárias – Universidade Estadual Paulista, *campus*, Jaboticabal (SP), Jaboticabal (SP).

Coelho, W. M.; Amarante, A. F. T.; Apolinário, J. C.; Coelho, N. M. D.; Lima, V. M. F.; Perri, S. H. V.; Bresciani, K. D. S. Seroepidemiology of *Toxoplasma gondii*, *Neospora coninum*, and *Leishmania spp.* Infections and risck factors for cats from Brazil. *Parasitology Research*, New York v. 109, n. 1009-1013, 2011.

Corrêa, W. M.; Corrêa, C. N. M. Toxoplasmose. In:________. *Enfermidades Infecciosas dos Mamíferos Domésticos*. Rio de Janeiro: Medsi, 1992. p. 757-766.

Cruz, M. A.; Ullman, L. S.; Montaño, P. Y.; Hoffman, L.; Langoni, H.; Biondo, A. W. Seroprevalence of *Toxoplasma gondii* in cats from Curitiba, Paraná, Brasil. *Revista Brasileira de Parasitologia Veterinária*, Jaboticabal, v. 20, n. 3, p. 256-258, 2011.

De Feo, M.L.; Dubey, J.P.; Mather, T.N.; Rhodes, R.C. Epidemiologic investigation of seroprevalence of antibodies to *Toxoplasma gondii* in cats and rodents. *American Journal of Veterinary Research,* Schumburg, n. 63, v. 12, p. 1714-1717, 2002.

Dorny, P.; Speybroeck, N.; Verstraete, S.; De Becker, A.; Berkevens, D.; Vercruysse, J. Serological survery of *Toxoplasma gondii*, feline immunodeficiency virus and feline leukaemia virus in urban stray cats in Belgium. *Veterinary Record*, London, v. 151, 626-629, 2002.

Dubey, J. P.; Lappin, M. R. Toxoplasmosis And Neosporosis. In: Greene, C. E. *Infectious diseases of the dog and cat.* 2.ed. Philadelphia: WB Saunders Company, 1998. Cap. 90, p.493-503.

Dubey, J.P. Pathogenicity and infectivity of *Toxoplasma gondii* oocysts for rats. *Journal Parastiology*, Lawrence, v. 82, p. 951-956, 1996.

Dubey, J.P. Tachyzoite-induced life cycle of *Toxoplasma gondii* in cats. *Journal Parastiology*, Lawrence, v. 88, n. 4, p, 713-717, 2002.

Dubey, J. P. Toxoplasmosis. *Journal of the American Hospital Association*, Schaumburg, v. 189, n. 2, p. 166-170, 1986.

Dubey, J.P. Toxoplasmosis – a waterborn zoonosis. *Veterinary Parasitology*, Amesterdam, v.126, p. 57-72, 2004.

Dubey, J.P.; Graham, D.H.; Silva, D.S.; Lehmann, T.;Bahia-Oliveira, L.M.G. *Toxoplasma gondii* isolates of free range chickens from Rio de Janeiro, Brazil: mouse mortality, genotype, and oocist shedding by cats. *Journal Parastiology*, Lawrence, v. 89, p. 851-853, 2002.

Dubey, J.P.; Greene, C.E.; Lappin, M.R. Toxoplasmosis and neosporosis. In: Greene, C. E. Infectious diseases of the dog and cat. Philadelphia. W. B. Saunders, p. 818-834, 1990.

Dubey, J.P.; Hoover, E.A. Attempted Transmission of *Toxoplasma gondii* infection from pregnant cats to their kittens *Journal of the American Hospital Association,* Schaumburg, v. 170, n. 5, p. 538-40, 1977.

Dubey, J. P.; Mattix, M. E.; Lipscomb, T.P. *Lesions of neonatally induced toxoplasmosis in cats. Veterinary Patologhy, v. 33, p. 290-5, 1996.*

Dubey, J. P.; Carpenter, J. L. Histologically confirmed clinical toxoplasmosis in cats: 100 cases (1952-1990). *Journal of the American Hospital Association*, Schaumburg, v. 203, n. 11, p. 1556-1566, 1993.

Dubey, J.P.; Lappin, M. R. Toxoplasmosis and neosporosis. In: Greene C. E. Infectious diseases of the dog and cat. 3. ed. St. Louis: Elsevier, 2006, p. 754-775.

Dubey J. P.; Powell, M. Ante-mortem diagnosis diarrhea, oocyst shedding, treatment, isolation and genetic typing of *Toxoplasma gondii* associated with clinical toxoplasmosis in a naturally infected cat. *The Journal Parasitology*, Lawrence, v.5, n. 58, p. 2012.

Frenkel, J.K. Toxoplasmosis: parasite, life cycle, pathology and immunology. In: Hammond, D.M.; Long, P.L. The Coccidia : eimeria, isospora, toxoplasma and related genera. Baltimore. University Park Press. 1973. p. 343-410

Frenkel, J.K.; Dubey, J.P.; Miller, N.L. *Toxoplasma gondii* in cats: fecal stages identified as coccidian oocysts. *Science,* Washington, v.167, n. 919, p.893-896, 1970.

Gauss C.B.L.; Almería S.; Ortuño A.; Garcia F.; Dubey J.P. Seroprevalence of *Toxoplasma gondii* antibodies in domestic cats from Barcelona, Spain. *Journal Parastiology*, Lawrence, v. 89, n. 5 , p. 1067-1068, 2003.

Garcia, J.L.; Navarro, I.T.; Ogawa, L.; De Oliveira, R.C. Soroepidemiologia da toxoplasmose em gatos e cães de propriedades rurais do municipio de Jaguapitã, Estado do Paraná, Brasil. *Ciência Rural*, v. 29, p. 99-104, 1999.

Harvey, A. M.; Greeffydd-Jones, T. J. Feline Inflamatory liver disease. In: Ettinger, S.J.; Feldman E.C. eds. Textbook of veterinary internal medicine – diseases of the dog and the cat. 7 ed. St. Louis, Minissouri: Elservier Sounders, 2010, p. 1643-1648.

Hutchison, W.M.; Dunachie, J.F.; Work, K. The fecal transmission of *Toxoplasma gondii. Acta Pathologica, Microbiologica et Immunologica*, Scandinavia, v.74, N.3., p.462-4, 1968.

Hutchison, W.M.; Dunachie, J.F.; Siim, J.C.; Work, K. Coccidian - like nature of *Toxoplasma gondii. Journal Britisch Medical,* Oxford, v.1, p.142-4, 1970.

Hutchison, W.M., Dunachie, J.F., Work, K. Siim, J.C. The life cycle of the coccidian parasite *Toxoplasma gondii* in the domestic cat. *Transations Royal Society Tropical Medicine and Hyg*iene, Oxford, v.65, n.3, p.380-399, 1971.

Langoni, H.; Silva, A.V. Da; Cabral, K. De G.; Cunha, E.L.P.; Cutolo, A.A. Prevalence of toxoplasmosis in cats from Sao Paulo and Parana States. *Brazilian Journal of Veterinary Research Animal Science*, São Paulo, 38, n. 1/6, p. 243-244, 2001.

Lappin, M. R. Feline Toxoplasmosis raises health concerns for humans. DVM newsmagazine, june 2002. Disponível em: http://veterinarynews.dvm360.com/dvm/data/articlestandard//dvm/392002/32688/article.pdf Acesso em 20 de Agosto de 2012.

Lappin, M. R. et al. Segredo em Medicina Interna de Felinos, Porto Alegre: Armed, 2004.

Lucas, S. R. R., Hagiwara, M. K.; Reche, A. Germano, P. M. L. Ocurrence of antibodies to *Toxoplasma* in cats naturally infected with immunodeficiency virus. Brazilian *Journal of Veterinary Research Animal Science*, São Paulo, v. 35, N. 1, p. 41-45, 1998.

Meireles, L.R.; Galisteo,.A.J.Jr.; Pompeu, E.; Andrade, H.F. Jr. *Toxoplasma gondii* spreading in an urban area evaluated by seroprevalence in free-living cats and dogs. *Tropical Medicine Internal Health,* Oxford, v. 9, n. 8, p. 876-881, 2004.

Miró, M.; Montoya, A.; Jiménez, S.; Frisuelos, C.; Mateo, M.; Fuentes, I. Prevalence of antibodies to *Toxoplasma gondii* and intestinal parasites in stray, farm and household cats in Spain. *Veterinary Parasitology*, Amesterdam v. 126, p. 249-255, 2004.

Monteiro, S. G. Parasitologia na Medicina Veterinária, São Paulo: Roca, 2010.

Omata, Y.; Oikawa, H.; Kanda, M.; Mikazuki, K.; Nakabayashi, T.; Suzuki, N. Experimental feline toxoplasmosis: humoral immune responses of cats inoculated orally with *Toxoplasma gondii* cysts and oocysts. *The Journal Veterinary Science*, Tokio, v. 52, n. 4, p. 865-867, 1990.

Park, C. H.; Ikadai, E.; Yoshida, H. Isomura, H.; Inukai, H.; Oyamada, T. Cutaneous toxoplasmosis in a female japanese cat. *Veterinary Pathology*, Amsterdam, v. 44, n. 5, p. 683-687, 2007.

Pena, H.F.J. Isolamento e caracterização biológica e genotípica de *Toxoplasma gondii* (Nicolle e Manceux, 1909) de gatos do estado de São Paulo. 2004. 126 f. Tese (Doutorado em Medicina Veterinária) – Faculdade de Zootecnia e Medicina Veterinária, Universidade de São Paulo, São Paulo, 2004.

Papich, M. G. *Manual Saunders: Terapêutico Veterinário*. 2 ed. São Paulo: MedVet, 2009, 774p.

Powell, C.C.; Brewer, M.; Lappin, M.R. Detection of *Toxoplasma gondii* in the milk of experimentally infected lactating cats. *Vet. Parasitol.,* v.102, n.1, p.29-33, 2001.

Sato, K.; Iwamoto, I.; Yoshike, K. Experimental toxoplasmosis in pregnant cats. *The Journal Veterinary Science*, Tokio v. 55, n. 6, p. 1005-1009, 1993.

Sakamoto, C. A. M.; Costa, A. J.; Gennari, S. M.; Pena, H. F.; Toniollo, G. H.; Lopes, W. D.; Bichuette, M. A.; Betini, M. A.; Amarante, A. F.; Bresciani, K. D. S.. Experimental infection of pregnant queens with two major Brazilian clonal lineages of *Toxoplasma gondii. Parasitology Research*, New York, v. 105, p. 1311-1316, 2009.

Silva, J. C. R.; Gennari, S. M.; Ragozo, A. M. A.; Amajones, V. R.; Magnabosco, C.; Yai, L. E. O.; Ferreira-Neto, J. S.; Dubey, J. P. Prevalence of *Toxoplasma gondii* antibodies in sera of domestic cats from Guarulhos and São Paulo, Brazil. *Journal Parastiology*, Lawrence, v. 88, n. 2, p. 419-20, 2002.

Summer, B.; Achland, M.L. *Toxoplasma gondii* antibody in domestic cats in Melbourne. *Australian Veterinary Journal,* Oxford, v. 77, p. 447 – 9, 1999.

Smart, M. E.; Downey, R. S.; Stockdale, P.H. Toxoplasmosis in cat associated with cholangitis and progressive pancreatitis. *The Canadian Veterinary Journal*, Ottawa, v.14, n. 122, p. 313-316, 1973.

Sobrinho, L. S.; Rossi, C. N.; Vides, J. P.; Braga, E.T.; Gomes, A. A.; De Lima, V. M.; Generoso, D.; Langoni, H.; Leutenegger, C.; Biondo, A. W.; Laurenti, M. D.; Marcondes, M. Coinfection of *Leishmania chagasi* with *Toxoplasma gondii*, Feline Immunodeficiency Vírus (FIV) and Feline Leukemia Vírus (FeLV) in cats from an endemic area of zoonotic visceral leishmaniasis. *Veterinary Parasitology*, Amsterdam, v. 187, n. 1-2, p. 302-306, 2012.

Swinger, R. L.; Shmidt, K.A.; Dubielzig, R. R. Keratoconjunctivitis associated with *Toxoplasma gondii* in a dog. *Veterinary Ophthalmology*, California, v. 12, n. 1, p. 56-60, 2009.

Urquhart, G.M.; Armour, J.; Duncan, J.L.; Dunn, A .M.; Jennings, F.W. Parasitologia Veterinária. 2. ed. Rio de Janeiro: Guanabara Koogan, 1998. 273 p.

Witte, H. M.; Piekarski, G. Die Oocysten-Ausscheidung bei experimentell infizierten Katzen in Abhängigkeit vom Toxoplasma-Stamm, *Zeitschrift für Parasitenkd*, Berlim, v. 33, n.4, p. 358-60, 1970.

Zhang, S.; Wei, M. X.; Ding, Z. Y.; Xu, X. P. Comparison a modified agglutination test (MAT), IHAT and ELISA for detecting antibodies to *Toxoplasma gondii. Acta Parasitologica Medica Entomologica Sinica*, China, v. 8, n. 4, p. 199-203, 2001.

In: *Toxoplasma Gondii* ISBN: 978-1-62618-349-0
Editors: M. Goodson and D. Winslow 

*Chapter VII*

# Toxoplasmosis in Public Health

***Katia Denise Saraiva Bresciani***[1,2*]**,**
***André Luiz Baptista Galvão***[2]**,**
***Amanda Leal de Vasconcellos***[2]**,**
***Rodrigo Rabelo dos Santos***[2]**,** ***Luiz Silveira Neto***[2]**,**
***Sandra Valéria Inácio***[1]**,** ***José Antonio Soares***[3]**,**
***Jancarlo Ferreira Gomes***[4]
***and Thais Rabelo dos Santos***[1,2]

[1]Faculdade de Medicina Veterinária de Araçatuba,
UNESP, Araçatuba, São Paulo, Brazil
[2]Faculdade de Ciências Agrárias e Veterinárias, UNESP,
Jaboticabal, São Paulo, Brazil
[3]Faculdades Integradas de Itapetininga, FKB, Itapetininga,
São Paulo, Brazil
[4]Instituto de Biologia e Instituto de Computação da Universidade de
Campinas, São Paulo, Brazil

[*] Paulo. bresciani@fmva.unesp.br.

# Abstract

Infection by *Toxoplasma gondii* is particularly interesting from the point of view of public health. Attention should be drawn to the risk factors for acquiring this prenatal infection, considering its pathogenesis and sequelae. Women in initial stages of gestation may undergo miscarriage, premature birth, neonatal death and even the classical Sabin's Triad, characterized by retinochoroiditis, cerebral calcifications, hydrocephaly or microcephaly. The risk of acquiring this disease is higher in the postnatal life, when severe behavioral changes such as attention deficit and schizophrenia may occur, reducing the life quality of individuals. There is strong association of reactivation of this disease in cases of acquired immunodeficiency syndrome – AIDS, commonly leading to secondary infection of the central nervous system in immunocompromised people and severe encephalitis. The aim of this chapter wil be discuss educational programs directed at reducing environmental contamination by *T. gondii* would eventually decrease the cost of treating humans with clinical toxoplasmosis.

# Importance of Toxoplasmosis in Public Health

## Introduction

Toxoplasmosis, a zoonosis of worldwide distribution, has importance in Human and Veterinary Medicine. Literature references that this parasite can be present in more than 30 species of birds and 300 mammals (Rey, 2008).

This coccidian causes behavioral changes such as schizophrenia, attention deficit disorder and obsessive compulsive disorder (Lafferty, 2005). During pregnancy, the congenital form of the disease can occur (Bachmeyer et al., 2006; Hung et al., 2007), also being observed cases of antenatal depression and anxiety, but its mechanism of action is not fully elucidated (Groer et al., 2011).

Many cases of reactivation of this disease are associated with Acquired Immunodeficiency Syndrome - AIDS (Bachmeyer et al., 2006; Hung et al., 2007), especially with manifestation of severe encephalitis (Pradhan et al., 2007, Jones and Dubey, 2012). Educational programs focused on reducing reducing environmental contamination by *Toxoplasma gondii* would reduce the cost of human treatment with clinical toxoplasmosis (Santos et al., 2010, Jones and Dubey et al., 2012).

## Human Toxoplasmosis

The causative agent of this disease is the *T. gondii* parasite described almost simultaneously in rabbits, in 1908, by Splendore, in Brazil and Nicolle and Manceaux in *Ctenodactylus gondi,* a rodent from the north desert of Africa (Kawazoe, 2002).

In 1923, Jankü observed the agent in the eye of a 11 months child, in Prague, Czechoslovakia and Torres, in 1927, in Rio de Janeiro, incriminated the *T. gondii* as the cause of congenital meningoencephalitis, myocarditis and myositis in a newborn with 29 days of life.In both works, the parasite was identified as *Encephalitozoon* (protozoan of difficult distinction from *Toxoplasma*, without special stains).In 1937, Wolf and Cowan, reported the disease in many children and evidenced transplacental transmission.

The frequency of infection began to be characterized from 1948, with Sabin and Feldman, and then with other researchers, while developing the first diagnostic evidences of the disease. In the 40s, Pinkerton and Weinman (1940) and Pinkerton and Henderson (1941) reported acute toxoplasmosis in adults (Amato Neto et al., 1995). Coccidian protozoan obligate intracellular, belonging to the Kingdom Protista, Sub-kingdom Protozoa, Phylum Apicomplexa, Genus *Toxoplasma* and species *T.gondii* (Kawazoe, 2002).

Studies indicate that 10% to 20% of human infections with *T. gondii* occur in adults with symptomatic evolution. The most common clinical manifestation in the acute phase of this zoonosis is lymphadenopathy, usually in the cervical area, although other groups of nodes may be enlarged. There may also be fever, malaise, night sweats, myalgia, maculopapular rash and lymphocytic atypia (Bonametti et al., 1997).

In humans, according to seroepidemiologic surveys for *T. gondii*, it was found that antibody titers increased with age in both sexes and varies according to climatic factors, dietary and/or hygienic habits (Sáfadi, 2000).

Brazil presents human indexes that are among the highest reported, where recorded serosurveys showed a prevalence ranging from 30.34 to 97.10% (Table 1) and it was found that for every 1,000 live births, five presented the disease (Castilho, 1976). It is estimated that around 70% of the population has already been infected at some point of life (Vergara et al., 1985). In the state of Rio Grande do Sul, 140,914 newborns were analyzed, being possible to determine a prevalence of one for each 3,000 live births (Neto et al., 2000). n Campos de Goytacazes, in the state of Rio de Janeiro, the statistics are even higher, with five cases of congenital toxoplasmosis for each 2,550 newborns (Petersen et al., 2001).

**Table 1. Values of occurrence (%) of *anti-Toxoplasma gondii* antibodies in humans from different brazilian states, sources, techniques and respective authors**

| Region | State | Reference | Technique | Occurrence (%) |
|---|---|---|---|---|
| North | Pará | Costa (2000) | IFA | 85.00 |
| | | Carmo et al. (2004) | - ELISA IgG | 82.93 |
| | | | ELISA - IgM | 0.00 |
| Northeast | Pernambuco | Coelho et al. (2003) | ELISA | 75.00 |
| | Ceará | Rey and Ramalho (1999) | ELISA | 53.10 |
| South | Paraná | Soccol et al. (2003) | IFA - IgG | 43.40 |
| | | Garcia et al. (1999) | IFA | 65.80 |
| | | Daguer et al. (2004) | IFA | 67.20 |
| | | | ELISA | 84.40 |
| | | Lopes et al. (2005) | IFA - IgG | 50.40 |
| | | Millar et al. (2007) | IFA - IgG | 48.10 |
| | | | IgG-ELISA | 58.60 |
| | | | IFA - IgM | 0.00 |
| | | Gonçalves et al. (2006) | IFA | 70.00 |
| | | Giraldi et al. 2002 | IFA | 43.00 |
| | Rio Grande do Sul | Detanico and Basso (2006) | ELFA | 36.80 |
| Southeast | São Paulo | Ishizuka (1978) | IFA | 97.10 |
| | | Olbrich-Neto and Meira (2004) | IgG | 60.00 |
| | | | IgM | 2.10 |
| | | Silveira (2002) | IFA | 41.66 |
| | Rio de Janeiro | Uchoa et al. (1999) | IFA | 69.90 |
| | | | ELISA | 78.64 |
| | | Souza et al. (1995) | IFA | 84.70 |
| | | Coutinho et al. (1981) | IFA | 78.70 |
| | Minas Gerais | Segundo et al. (2004) | ELISA | 51.60 |
| Central West | Goiás | Avelino et al. (2004) | IFA | 51.20 |
| | Mato Grosso do Sul | Araujo et al. (2000) | IHA | 30.34 |
| | | Figueiró-Filho et al. (2005) | ELISA | 92.00 |
| | Mato Grosso | Amendoeira et al. (2003) | IFA | 80.40 |
| | | Santos et al. (2010) | IFA - IgG | 97.41 |

IFA: Indirect Immunofluorescence Assay; ELISA: Enzyme Linked Immunosorbent Assay; ELFA: Enzyme Linked Fluorescent Assay; IHA: Indirect Hemagglutination Assay.

The intermediate hosts, humans and animals, can acquire the disease by eating raw or undercooked meat, infected with tissue cysts, as well by the consumption of raw vegetables and water contaminated with oocysts (Garcia, 2007). Tachyzoites of *T. gondii* can be acquired by consuming unpasteurized milk, eggs, blood transfusion, semen, organ transplantation, laboratory accidents with biological material and by transplacental via (Dubey, 1986; Dubey et al., 1990; Dubey, 1994; Bonametti et al., 1997; Tenter et al.,2000; Powell et al., 2001; Kawazoe, 2002; Rey, 2008).

Sposito-Filha et al. (1992) isolated 20 strains of *Toxoplasma* from the diaphragm muscle of 136 sheep.Da Silva and Langoni (2001) identified this protozoan in the brain and diaphragm of 34 of 40 sheep serologically reactive to indirect immunofluorescence. These works of viable parasites isolation in edible tissues, as well in goat milk (Chiari and Neves, 1984) and sheep milk (Santos, 2012), reveal the importance of these sources of infection for humans.

The human toxoplasmosis develops only in situations of immunocompromise due, greatly, to the Human Immunodeficiency Virus (HIV), what grant to the disease a prominent place as cause of deaths from infections considered opportunistic. In addition, other diseases/conditions could be associated with this parasite, such as: lymphoma, leukemia, organ transplants or prolonged use of cytostatics and corticosteroids (Frenkel; Bermudez, 2006).

This coccidia is present in nature in three evolutionary forms: oocysts, which form up to four sporozoites inside; tachyzoites, proliferative form and tissue cysts containing bradyzoites (Dubey et al., 1998; Garcia, 2007).

Although most cats who had contact with *T. gondii* develop immunity and do not shed more oocysts, Dubey and Frenkel(1974) proved that seropositive cats, facing immunosuppression cases, may restart shedding oocysts. Subsequently, Dubey (1995) reinfected seropositive cats, six years after seroconvertion and 55% of them restarted shedding oocysts in the feces, however, in amounts below those of the primary infection.The cats are probably the main sources of environmental contamination, because, when infected, release large quantities of oocysts, allowing the continuity of the parasite biological cycle. It is estimated that only in one evacuation about 10 to 100,000 oocysts per gram of faeces can be eliminated (Tenter et al., 2000; Rey, 2008).

Tenter et al. (2000) observed that in islets and atolls that were never inhabited by cats, there are no infections by this protozoan. Often, the direct contact with domestic cat is considered the main cause of toxoplasmosis transmission in humans, but this is an incorrect information. Epidemiological studies conducted in various locations around the world, using different methods show serological prevalence of antibodies to *T. gondii* between 3% and 88% of the evaluated cats. However, in sectional studies, only between 0 and 4.3% of cats eliminate oocysts, although this occurrence may reach 34% in specific regions, but this data does not reflect a global reality (Dabritz and Conrad, 2010).

If the frequency of oocysts shedding by infected cats is low, why is toxoplasmosis in humans so common? A cat can eliminate from three to 810 million oocysts per infection (Dabritz and Conrad, 2010). Direct contact with this host is irrelevant in the epidemiological chain of the disease due to their habit of licking, and this promotes a mechanical removal of oocysts, preventing sporulation in its pelage. So the environmental control becomes more relevant and effective in preventing the occurrence of this disease in humans, once the oocysts remain viable in the soil for up to 18 months, depending on the temperature, humidity and exposure to sunlight (Frenkel et al., 1975; Yilmaz and Hopkins, 1972) and between six and 54 months in water (Dubeyet al., 1998) or oceans (Lindsay and Dubey, 2009).

Our research group noted that *T. gondii* oocysts are widely distributed on the soil of elementary public schools in our region, likely constituting the main contamination source for these children (Santos et al., 2010).

Behavioral changes, such as attention-deficit schizophrenia (Lafferty, 2005) are possibly a result of high concentrations of dopamine in the limbic area of the brain due to the presence of {1}*T. gondii*{/1} (Yolken et al., 2001). These disorders are considered more common in young individuals (under age 38), because they are impulsive and less careful with their hygiene, proper washing of fruits, vegetables and cooking of meat products, thus, having a higher risk of acquiring this disease (Okusaga et al., 2011).

In the first infection, during pregnancy, about 90% of cases are asymptomatic or oligosymptomatic.The tachyzoite, reaches the fetus by transplacental via, causing damage of diferent levels of severity, depending on the virulence and/or lineage, on the strain of the mother's immune response capacity and on the gestational period in which the woman is, that can result, even, in fetal death or severe clinical manifestations (Groër et al., 2011).

Toxoplasmosis may go unnoticed at birth of an individual, however, it may manifest months or even years later in children born with this parasitosis. In these cases, the most common manifestations are retinocorioidite, cerebral calcifications, neuropathies and microcephaly, in up to 90%, 69%, 60% and 50% of the cases, respectively.In more severe cases of congenital infection, the newborn may show changes in cranial volume, intracranial calcifications and/or convulsions (Groër et al., 2011).

In the United States, toxoplasmosis is the second major foodborne illness, being responsible for hospitalization and even death, both in urban and rural areas (Dubey et al., 1998; Jones and Dubey, 2012).

Roghmann et al. (1999) observed low incidence of infection by *T. gondii* in vegetarian subjects (24%) when compared to carnivores (50%).Meat products from ovine and swine are more contaminated with this parasite cysts, and these are rarely found in bovine (Bonametti et al., 1997; Jones and Dubey, 2012).

In Spain, some people eat horse meat and García-Bocanegra et al. (2012) revealed that many horses from various regions of the country are serologically positive for toxoplasmosis.The demonstration of the occurrence of human toxoplasmosis associated to the milk ingestion from proven infected goats, is particularly important in rural and peri-urban areas where these animals are raised for the purpose of production of milk for domestic consumption (Chiari and Neves, 1984)

Only complete water treatment systems are capable of retaining most of *T. gondii* oocysts in steps of coagulation, sedimentation and filtration.Chlorination of water used for human consumption is not capable to inactivate them (Almeida et al., 2011). Contamination of municipal water reservoirs with a significant amount of *T. gondii* oocysts can lead to epidemics, as recorded in Canada, in 1995, or trigger outbreaks like the one reported in 2001, in Santa Isabel do Ivaí, Paraná State, Brazil (Bowie et al., 1997; Funasa, 2002; Almeida et al., 2011).

## Prevention and Control

In relation to the resistance of the cysts in meat, it is believed that they do not resist the processes of salting and heating (above 55°C for more than two minutes) used in the preparing of processed meat (Dubey, 1994).However, other studies show that the process of seasoning and sausages and salami, mostly, or cooled meat, may represent an important route of transmission of toxoplasmosis and may also be responsible for the high prevalence of toxoplasmosis eye injuries in southern Brazil (Glasner et al., 1992, Arias et al., 1996).

Humans and animals should not consume raw or undercooked meat, unpasteurized or not boiled dairy products, water without being treated and poorly washed fruits and vegetables (Tenter et al., 2000; Monteiro, 2010).

As daily routine, the cleaning of catteries, with removal of fecal material is indicated, since, in dependence on the environmental conditions, the *T.gondii* oocysts eliminated by cats, require at least 24 hours to sporulate and become infectious.It is also important to cover or encircle sand tanks for children recreation, especially when they are not in use, to prevent the access of cats (Monteiro, 2010; Santos et al., 2010).

Farm foods must be stored in locations without access to cats and mice . It is not also recommended the administration of leftovers to pigs (Taylor et al., 2010).

Dishes and utensils should be washed with hot water after contact with unwashed fruits and vegetables, unpasteurized milk and seafood.Pregnant women and immunosuppressed individuals need to wear gloves to get contact with the soil or dumps, avoiding possible touches in cat feces (Jones and Dubey, 2012).

Fetal infection could be attenuated or prevented by maternal treatment, especially after the effective clinical preventive diagnosis, parasitological, immunological and neurological. To do so, guidance on the transmission of toxoplasmosis in public medical clinics must be promoted (Jones and Dubey, 2012) as well as pay attention to the feline population control (Wang et al. 2012, Dubey et al., 2012). 2012, Dubey et al., 2012).

# Final Considerations

In our studies, thirty municipal elementary schools from Araçatuba were visited and 85 teachers were interviewed. The participation of cats in toxoplasmosis transmission was known by 92.94% (79/85), of which 82.35% (70/85) did not know the transmission routes. The dog was considered the disseminator of this disease by 80.00% (68/85) interviewees, and only 4.71% (4/85) cited ingestion of meat products as a route of *T. gondii* transmission, while 67.06% (57/85) did not know about this issue.

A survey that investigated teachers' knowledge degree about the major parasitical diseases was realized and, after lectures was verified the assimilation of the presented content. Interviews about toxoplasmosis were done in 60 of 18 educators from local elementary schools form Araçatuba, SP.

After this phase, short lectures on infections by protozoa were delivered. After the lectures, the same referring questions were repeated to the subject. Among the interviewed teachers for the survey, regarding knowledge on the major toxoplasmosis transmission ways, 86.66% incriminated felines, 70.00% quoted pidgeons, only one teacher mentioned about meat products and 28.33% did not know anything about the topic. After the lecture, as ways of transmission, 93.33% understood cats' role regarding the disease and 96.66% understood the topic concerning meat products (Cardia et al., 2007).

One hundred thirty four elderly people, from the institutions, were interviewed, using individual questionnaires. In relation to toxoplasmosis, 78.36% (105/134) did not know the meaning of the term, and 86.57% (116/134) were ignorant of the dissemination ways (Lima et al. 2008).

Based on the obtained data, it is possible to infer that there is the need of implanting a community education program directed to the improvement of basic concepts of toxoplasmosis control and prevention (tomé et al., 2005; Cardia et al., 2007; Lima et al., 2008; Jones and Dubey, 2012).

## References

Almeida, M. J. et al. Aspectos sociopolíticos da epidemiologia de toxoplasmose em Santa Isabel do Ivaí (PR). Ciência and Saúde Coletiva, v. 16, n. 1, p. 1363-1373, 2011.

Amendoeira MRR, Costa T, Spalding SM. Revista Souza Marques 1:15-35, 1999.

Bachmeyer, C.; Mouchnino, G.; Thulliez, P.; Blum, L. Congenital toxoplasmosis from an HIV-infected woman as a result of reactivation. *The Journal of Infection, London*, v.52, n.2, p.e55–e57, 2006.

Bonametti, A. M.; Passos, J. N.; Da Silva, E. M. K.; Bortoliero, A.L. Surto De Toxoplasmose aguda transmitida através da ingestão de carne crua de gado ovino. Revista da Sociedade Brasileira de Medicina Tropical, Uberaba, v. 30, n. 1, p. 21-25, 1997.

Bowie, W. R.; King, A. S.; Werker, D. H.; Issac-Renton, J. L.; Bell, A.; Eng, S. B.; Marion, S. A. Outbreak of toxoplasmosis associated with municipal drinking water. The BC Toxoplasma investigation Team. *Lancet*, London, v. 19, n. 350, p. 173-177, 1997.

Bresciani et al. Toxoplasmose Canina: Aspectos clínicos e patológicos. Semina: *Ciências Agrárias*, v. 29, n. 1, p. 189-202, 2008.

Cárdia, D. F. F.; Amarante, A. F. T. do; Bresciani, K. D. S.. Ponto de vista de professoras de escolas municipais do ensino fundamental de Araçatuba (SP) sobre as parasitoses. *Rev. Ciênc. Ext.* v.3, n.2, p.46, 2007.

Chiari, C.A.; Neves, D.P. Toxoplasmose humana adquirida através da ingestão de leite de cabra. Memórias do InstitutoOswaldo Cruz, Rio de Janeiro, n.3, v.79, p.337-340, 1984.

Coelho, W. M. D. et al. Seroepidemiology of Toxoplasma gondii, Neosporacaninum, and Leishmaniaspp. Infections and risk factors for cats from Brazil.*Parasitology Research*, v. 109, p. 1009-1013, 2011.

Dabritz, H.A.; Conrad, P.A. Cats and Toxoplasmosis: Implications for Public Health. *Zoonoses Public Health*, v. 57, 34-52, 2010.

Dubey J.P, Darrington C.; Tiao, N.; Ferreira, L.; Choudhary, S.; Molla B.; Saville, W.; Tilahun, G.; Kwok, O.; Gebreyes, G. Isolation of viable Toxoplasma gondii from tissues and feces of cats from addis Ababa, Ethiopia. *Journal Parasitology*, 2012 (in press).

Dubey, J. P. Toxoplasmosis. *Journal of the American Hospital Association*, Schaumburg, v. 189, n. 2, p. 166-170, 1986.

Dubey, J. P. Toxoplasmosis. *Journal of the American Veterinary Medical Association,* v. 205, n. 11, p. 1593-1598, 1994.

Dubey, J. P.; et al. Toxoplasmosis and neosporosis. In: Greene, C. E. Infectious diseases ofthe dog and cat.2.ed. Philadelphia: WB Saunders Company, 1998. Cap. 90, p.493-503.

Dubey, J.P. Duration of immunity to shedding of Toxoplasma gondii oocysts by cats. *Journal of Parasitology, Lawrence*, v.81, n.3, p.410-415, 1995.

Dubey, J.P.; Greene, C.E.; Lappin, M.R. Toxoplasmosis and neosporosis. In: Greene, C. E. Infectious diseases of the dog and cat. Philadelphia. W. B. Saunders, p. 818-834, 1990.

Frenkel, J. K.; A. Ruiz; M. Chinchilla. Soil survival of Toxoplasma oocysts in Kansas and Costa Rica. *Am. J. Trop. Med. Hyg.*,v. 24, 439-443, 1975.

Frenkel, J.K. Toxoplasmose. In: Veronesi, R.; Focaccia, R. Tratado de Infectologia. 2. ed. São Paulo: Atheneu, 2004. v. 2, p. 1310-1325.

Fundação Nacional de Saúde-Funasa Boletim Eletrônico Epidemiológico p.01-03, 2002. Disponível em: <http www.funasa com.br>

Garcia, L.C. Toxoplasma gondii. Diagnosis Medical Parasitology. 5th edition, A.S.M. Press, Washington, D.C.(USA). 2007, p 562-563.

García-Bocanegra, O. et al. Seroprevalence of Toxoplasma gondii in equids from Southern Spain. *Parasitology International,* v. 61, p. 421-424, 2012.

Groër, M. W. et al. Prenatal depression and anxiety in Toxoplasma gondii-positive women. *American Journal of Obstetrics andGinecology*, v. 204, p. 433. v.1, n.7, 2011.

Hung, C.C.; Fan, C.K.; Su, K.E.; Sung, F.C.; Chiou, H.Y.; Gil. V.; Ferreira, M.D.A. C.; Carvalho, J.M.; Cruz, C.; Lin, Y.K.; Tseng, L. F.; Sao, K.Y.; Chang, W. C.; Lan; H.S.; Chou, S.H. Serological screening and toxoplasmosis exposure factors among pregnant women in the Democratic Republic of Sao Tome and Principe. *Transactions of the Royal Society of Tropical Medicine and Hygiene*, v. 101, n.2, p. 134-139, 2007.

Jones, J. L. and Dubey, J. P. Foodborne Toxoplasmosis. *Clinical Infectious Diseases*, v. 18, p. 1-7. 2012.

Jones, J. L. andDubey, J. P. Foodborne Toxoplasmosis. *Clinical Infectious Diseases*, v. 18, p. 1-7. 2012.

Kawazoe , U. Toxoplasma gondii. In: Neves D. P. Parasitologia Humana. 10 ed. São Paulo: Atheneu, 2002; p. 147-156.

Lafferty, K. D. Look what the cat dragged in: do parasites contribute to human cultural diversity? Behavioural Processes, Amsterdam, v.68, n.3, p.279-282, 2005.

Lima, F.F., Koivisto, M.B., Perri, S.H.V., Bresciani, K.D.S. O conhecimento de idosos sobre parasitoses em instituições não governamentais do Município de Araçatuba, SP. Revista Ciencia em Extensão, v.4, n.1, p. 77-85, 2008.

Lindsay, D.S.; Dubey, J.P. Long-term survival of Toxoplasma gondii sporulated oocysts in seawater. *J. Parasitol.,* v. 95, 1019-1020, 2009.

Miman, O. et al. Is there any role of Toxoplasma gondiiin the etiology of obsessive-compulsive disorder? *Psychiatry Research*, v. 177, p. 263-265, 2010.

Miró, M.; Montoya, A.; Jiménez, S.; Frisuelos, C.; Mateo, M.; Fuentes, I. Prevalence of antibodies to Toxoplasma gondii and intestinal parasites in stray, farm and household cats in Spain. *Veterinary Parasitology*, Amesterdam v. 126, p. 249-255, 2004.

Monteiro, S.. G. Parasitologia na Medicina Veterinária, São Paulo: Roca, 2010.

Okusaga, O. et al. Toxoplasma gondiiantibody titers and history of suicide attempts in patients with schizophrenia. *Schizophrenia Research*, v. 133, p. 150-155, 2011.

Powell, C.C.; Brewer, M.; Lappin, M.R. Detection of Toxoplasma gondii in the milk of experimentally infected lactating cats. *Vet. Parasitol.,* v.102, n.1, p.29-33, 2001.

Pradhan, S.; Yadav, R.; Mishra, V. N. Toxoplasma meningoencephalitis in HIV-seronegative patients: clinical patterns, imaging features and treatment outcome. *Transactions of the Royal Society of Tropical Medicine and Hygiene*, London, v.101, n.1, p.25-33, 2007.

Rey, L. Toxoplasma gondii Toxoplasmose. Parasitologia. 4 ed. Rio de Janeiro: Guanabara Koogan, 2008, p 192-206.

Roghmann, M. C.; Faulkner, C. T.; Lefkowitz, A.; Patton, S.; Zimmerman, J.; Morris, J. G. Decreased seroprevalence for toxoplasma gondii in seventh day Adventists in Maryland. *The Amercian Journal of Tropical Medicine and Higiene*, v. 60, n. 5, p. 790-792, 1999.

Sáfadi, M . A. P. Toxoplasmose. Pediatria Moderna, São Paulo, v. 36, n. 1-2; p. 7-23, 2000.

Sakamoto, C. A. M.; Costa, A. J.; Gennari, S. M.; Pena, H. F.; Toniollo, G. H.; Lopes, W. D.; Bichuette, M. A.; Betini, M. A.; Amarante, A. F.; Bresciani, K. D. S.. Experimental infection of pregnant queens with two major Brazilian clonal lineages of Toxoplasma gondii. *Parasitology Research*, New York, v. 105, p. 1311-1316, 2009.

Sibley, L. D.; Khan, A.; Ajioka, J. W.; Rosenthal, B. M. Genetic diversity of Toxoplasma gondii in animals and humans. *Philosophical Transactions of the Royal Society Biological Sciences*, London, v. 364, n. 1530, 5749-2761, 2009.

Spalding, S. M. et al. Estudo prospectivo de gestantes e seus bebês com risco de transmissão de toxoplasmose congênita em município do Rio Grande do Sul. *Revista da Sociedade Brasileira de Medicina Tropical*, v. 36, n. 4, p. 483-491, 2003.

Sposito Filha, E. et al. Toxoplasma gondiiem ovinos: Isolamento do parasita a partir de diafragmas de animais procedentes do Estado do Rio Grande do Sul e abatidos em matadouros de São Paulo, para consumo humano. Revista Brasileirade Parasitologia Veterinária, São Paulo, n.2, v.1., p.118-119, 1992.

Taylor MA. Parasitologia veterinária, 3rd ed., Guanabara Koogan, Rio de Janeiro, 2010, 742 pp.

Tenter, A. M.; Heckeroth, A. R.; Weiss, L. M. Toxoplasma gondii: from animals to humans. *International Journal for Parasitology*; v.30, p.1217-1258, 2000.

Thais Rabelo dos Santos, Caris Maroni Nunes, Maria Cecilia Rui Luvizotto, Anderson Barbosa de Moura, Welber Daniel Zanetti Lopes, Alvimar Jose da Costa, Katia Denise Saraiva Bresciani. Detection of Toxoplasma gondii oocysts in environmental samples from public schools. *Veterinary Parasitology*, Vol. 171, pp. 53–57, 2010.

Tomé, R.O.; Serrano, A.C.M.; Nunes, C.M.; Perri, S.H.V.; Bresciani, K.D.S. Inquérito epidemiológico sobre conceitos de zoonoses parasitárias para professores de escolas municipais do ensino infantil de Araçatuba-SP. Revista Ciência em Extensão., v.2, n.1, p.38, 2005.

Tomé, R.O.; Serrano, A.C.M.; Nunes, C.M.; Perri, S.H.V.; Bresciani, K.D.S. Inquérito epidemiológico sobre conceitos de zoonoses parasitárias para professores de escolas municipais do ensino infantil de Araçatuba-SP. Revista Ciência em Extensão., v.2, n.1, p.38, 2005.

Wang*†, Q.; Jiang†, W.; Chen, YJ.; Liu, C.Y.; Shi, J.l.; Li, X.t. Prevalence of Toxoplasma gondii antibodies, circulating antigens and DNA in stray cats in Shanghai, China. Wang et al. *Parasites and Vectors*, p. 5:190, 2012.

Yilmaz, S. M.; S. H. Hopkins. Effects of different conditions on duration of infectivity of Toxoplasma gondii oocysts. *J. Parasitol.,* v. 58, 938-939, 1972.

Yolken, R. H.; Bachmann, S.; Rouslanova, I.; Lillehoj, E.; Ford, G.; Torrey, E. F.; Antibidoidies to Toxoplasma gondii in individuals with first-episode schrizophrenia. *Clinical Infection Disease*, v. 32, n. 4, p. 842-844, 2001

In: *Toxoplasma Gondii* ISBN: 978-1-62618-349-0
Editors: M. Goodson and D. Winslow 

*Chapter VIII*

# Control Program of Congenital Toxoplasmosis

***Italmar Teodorico Navarro[1], Regina Mitsuka Breganó[1], Fabiana Maria Ruiz Lopes Mori[2], Tercilia Oliveira Rodrigues[3], Thaís Rabelo dos Santos[3,4], Rodrigo Rabelo dos Santos[4] and Katia Denise Saraiva Bresciani[3,4*]***

[1]Centro de Ciências Biológicas, UEL, Londrina, Paraná
[2]Centro Universitário Filadelfia, UNIFIL, Londrina, Paraná
[3]Faculdade de Medicina Veterinária de Araçatuba, UNESP, Araçatuba, São Paulo, Brazil
[4]Faculdade de Ciências Agrárias e Veterinárias, UNESP, Jaboticabal, São Paulo, Brazil

## Abstract

Toxoplasmosis is a zoonotic disease caused by coccidia *Toxoplasma gondii,* that affects almost all homeothermic animal species, including birds, mammals and also humans. The implementation of control programs including preventive measures such as laboratory tests

* Email: bresciani@fmva.unesp.br

frequently promotes early diagnosis and appropriate indication of adequate antiparasitic treatment to pregnant women, and is, therefore, able to diminish the seriousness of the sequelae of toxoplasmosis in the fetus. Important aspects of the control of congenital toxoplasmosis programs for maternal and neonatal screening, the toxoplasmosis control program were reviewed.

# Introduction

Toxoplasmosis is a zoonotic disease caused by coccidia *Toxoplasma gondii,* obligate intracellular parasite, with a complex biological cycle and that affects almost all homeothermic animal species, including birds, mammals and also humans (Dubey and Beattie, 1988, Tenter et al., 2000).

The toxoplasmosis transmission ways can be varied between man and animals. Non-pasteurized goat's milk, eggs, blood transfusion and organ transplantation, are less epidemiologically important transmission routes (Dressen, 1990). After birth, the main ways of human infection is through the consumption of meat and its derivatives containing cysts in the muscle fibers, and also by eating vegetables, fruits or water contaminated with oocysts (Navarro et al., 1992).

In relation to the forms of infection, these vary according social and hygienic factors, environmental humidity, professional occupation, habits and customs of a given region. The percentage of people infected by *T. gondii* is higher in people who have the habit of eating raw or undercooked meat.This fact evidence that the cultural habit of a population is one of the factors that could explain the variability of this infection in different geographical areas of a country (Silveira et al., 1987).

In humans, transplacental transmission was first demonstrated by Wolf and Cowen (1937), Wolf et al. (1939) that contributed primarily to the beggining of the diagnosis of this protozoan, therefore, it is necessary to be alert to the risk factors for acquisition of prenatal infection as well to consider possible pathogenesis and sequelae (Ortega-Benito, 2001; Boyer et al., 2005; Bachmeyer et al., 2006; Hung et al., 2007).

Pinkerton and Weinman (1940) described, with pioneering, cases of toxoplasmosis in postnatal life (Silva et al., 2001; Giraldi et al., 2002; Escuissato et al., 2004). Although this disease is usually asymptomatic, it is estimated that one third or more of the human population has antibodies to the parasite (Montoya; Liesenfeld, 2004).

In the case of acutely infected pregnant women, congenital toxoplasmosis results from transplacental transmission of *T. gondii*, and when the mother is infected during the beggining of pregnancy, it can cause abortion or malformations (Dunn et al., 1999).Therefore, infected children may present, at birth, the classic signs and symptoms of congenital toxoplasmosis (hydrocephalus, chorioretinitis, mental retardation and cerebral calcification) or may be born normal and develop sequelae during childhood or, even, in adulthood (Curtain- Borja et al., 2010).

The implementation of control programs including preventive measures such as laboratory tests frequently promotes early diagnosis and appropriate indication of adequate antiparasitic treatment to pregnant women, and is, therefore, able to diminish the seriousness of the sequelae of toxoplasmosis in the fetus (Foulon et al., 1994; Hohlfeld et al., 1989; Dias et al., 2011.)

Given the above, and in order to highlight the importance of control programs for congenital toxoplasmosis, in this chapter are reviewed important aspects of the control of congenital toxoplasmosis, programs for maternal and neonatal screening, the toxoplasmosis control program in Brazil and Paraná state.

# Control of Congenital Toxoplasmosis

Mitsuka-Breganó et al., 2010 emphasize that the control of congenital toxoplasmosis consists of three prevention steps: primary, secondary and tertiary. Primary prevention is essential for infection control and becomes through guidelines on hygienic-dietary measures that pregnant women should take to prevent primary infection, thus preventing fetal infection. Secondary prevention is the serological diagnosis of maternal infection and treatment of pregnant women in order to prevent or reduce infection sequelae after maternal/fetal infection. The tertiary prevention intend to reduce the sequelae of children with congenital toxoplasmosis, during the first year of life, through treatment in the first year of life. Lopes-Mori et al. (2011) observed that public health systems in many countries, have not yet adopted specific and uniforms policies for the prevention of toxoplasmosis. Countries with a high incidence of infection implemented prenatal screening programs, as in France (Thulliez, 1992), Áustria (Aspock and Pollak, 1992) and Slovenia (Logar et al., 2002). Denmark (Lebech et al., 1999) and Poland (Paul et al., 2000) with low incidence, adopted neonatal screening. In countries where toxoplasmosis is rare, as in the UK (Gilbert and Peckham, 2002) and USA (Boyer et al., 2005), is not adopted any universal serologic screening program. Lappalainen et al. (1995) in Finland, where the prevalence in pregnant women is 20.3% and the incidence is 2.4 per 1,000, shows that the analysis of the cost and effectiveness of screening programs are important in the decision of public health policies. The prenatal screening associated with health education is economically viable when the incidence of maternal infection exceeds 1.1 per 1,000 pregnant women and this screening is recommended even for countries with low incidence due to the serious consequences of congenital toxoplasmosis. Thus, education programs are considered a powerful intervention due to its low risk and cost and because the mother is motivated to protect the health of her baby (Elsheikha, 2008). The prenatal screening is not mandatory in Brazil, but due to the high prevalence of maternal toxoplasmosis, it is recommended and offered for free in some locations in the states of Mato Grosso do Sul, Minas Gerais, Paraná, Rio Grande do Sul and Goiás (Mitsuka-Breganó et al., 2010).

Several Brazilian experts, after analyzing the prevalence and cost/benefit of implementing preventive actions, suggest the inclusion of toxoplasmosis in Neonatal Screening Program, complementary to maternal screening (Neto et al., 2004; Lago et al., 2007; Mitsuka-Breganó et al., 2010; Lopes-Mori et al., 2011). Among other reasons for this inclusion, there is the existence of some factors that complicate the prevention, particularly in cases where the pregnant woman acquires the infection after performing the last serology, because there are still women who do not make regular prenatal care. This complicates the diagnosis, prevention and treatment, since in this phase, which fetal transmission rate is higher, the infection could be detected (Mitsuka-Breganó et al. 2,010; Lopes-Mori, 2010), necessitating, therefore, a program that monitors and treat the pregnant woman throughout her pregnancy period, as well as the child.

## Health Surveillance Program of Congenital Toxoplasmosis in Pregnant Women Served by the Sistema Único De Saúde (SUS) in Londrina, PR

Given the obvious cost/benefit of including toxoplasmosis in Neonatal Screening Program, complementary to maternal screening (Lago et al., 2007;

Neto et al., 2004), a team of experts from various fields (Health Surveillance Department from Ministry of Health, Health Department of Paraná State and the University of Londrina) implemented the Program for Health Surveillance of Congenital Toxoplasmosis in Pregnant Women Served by the Sistema Único de Saúde (SUS) em Londrina, in 2006. This program established, tested and validated, in the public health service, protocols for diagnosis, treatment and behaviors of mothers and children.

The actions from this program include serological screening in the first prenatal consultation and quarterly serological monitoring, until guidance on infection prevention action for pregnant women. It is noteworthy that these protocols are listed in the manual: "Congenital Toxoplasmosis and Gestational: Manual for Health Surveillance, Diagnosis, Treatment and Actions" (Mitsuka-Breganó et al. 2010). Moreover, given the importance of this program and its success, the Ministry of Health has recommended the use of the same model to install surveillance of toxoplasmosis nationally.

After implantation and validation in the city of Londrina, the program was implemented in other cities in the state of Paraná, as Cascavel, Cambé, Rolândia, Palotina, Jesuítas and Umuarama as a way to evaluate the actions and in other social realities and ecosystems (Lopes- Mori et al., 2011, Bittencourt et al., 2012, Dias et al., 2011).

## Stages of Program Implementation in Londrina-PR

The implementation of the program in the city of Londrina-PR was organized as follows: First step - definition of screening and confirmatory tests, referral services, mandatory reporting, investigation of cases and drug delivery; Second stage - Workshops for training of professionals in public health; Third stage - implementation of the program in Basic Health Units, and; Fourth step - program evaluation (Mitsuka-Breganó et al., 2010; Lopes-Mori et al., 2011).

## First Step: Definition of Screening and Confirmatory Tests, Referral Services, Mandatory Reporting, Investigation of Cases and Drug Delivery

In order to define some important points for the implementation of the program, the professionals responsible for various service teams during pregnancy in the Health Department of the city of Londrina are gathered from: Municipal Epidemiological Surveillance, municipal laboratories, reference services for care to pregnant women and children and Health Regionals.

This team has the responsibility to make decisions about various factors that contribute to the smooth running of the program, such as: the methods of serological screening and confirmatory; definition of the referral clinics and also the routing, definition of the compulsory notification form, and the application of epidemiological investigation form and the powers of Epidemiological Surveillance and Health Surveillance of reported cases, and; the definition of the manner of the liberation of drugs for treating pregnant women and children (Mitsuka-Breganó, 2009).

## Second Step: Training Workshops

This step aims to conduct training workshops addressing aspects of the life cycle of the parasite, with emphasis on transmission and prevention, laboratory diagnostics and program actions to be implemented.

The workshops, conducted initially in person, taught by a team of veterinarians, doctors, nurses and biochemists at the University of Londrina are directed at three target groups that serve pregnant women and children: doctors (gynecologists, pediatricians, infectious disease specialists, ophthalmologists, neonatologists), nurses, servers of Sector of Epidemiological Surveillance of the City and FHP teams (Family Health Program); biochemist responsible for serological screening and confirmatory; Community Health workers responsible for conducting an active search for pregnants not present in serology tests and the actions of health education, with emphasis on preventive measures (Mitsuka-Breganó, 2009; Lopes-Mori et al., 2011).

The training is necessary due to the fact that the differentiation of protocol service in BHU (Basic Health Units) may delay the diagnosis of congenital toxoplasmosis, complicating the decision-making of the this disease prevention.

On the other hand it was observed that the implementation of programs at the municipal level as in regional or state level, some factors complicate constant training of the teams involved, among them we can highlight the possibilities of professionals participate in updating workshops, the availability of experts to minister the same and, finally, the distances between regional and local authorities. Also, due to the difficulty in reconciling the time of healthcare professionals for classroom trainment aiming updating, upgrading and notifications of new approaches and protocols, continuing education of health professionals end up facing the pressure of scientific progress, being driven to seek new educational approaches (Rodrigues et al., 2011).

In this context, distance education stands out as an innovative teaching method that effectively lowers the barriers of time and geographical space, enabling audiences to meet their specific targets that would hardly be served by classroom education (Santos; Market, 2010).

## Third Step: Beggining of Activities in Basic Health Units (BHU)

Once defined and set forth the goals of the earlier steps, it is started the implementing of the program in all BHU (Basic Health Units), under the coordination of an officer appointed by the Health Department of the Municipality and technical support specialists (Mitsuka- Breganó et al., 2009).

Health education campaigns are promoted in order to target pregnant women and their families, about the preventive measures through the supporting materials previously manufactured for the training and guidance to pregnant women (Manual, Leaflet, Folder, DVD, poster, etc.).

It is interesting and recommended, at this stage, to know the epidemiological characteristics of congenital toxoplasmosis and pregnancy in the city through the use of questionnaires to pregnant women attended in the UBS in the period (Mitsuka-Breganó, 2009).

## Fourth Step: Evaluation of the Program

The program was implemented in all 52 UBS which perform prenatal consultations (39 in urban and 13 in rural areas) in the city of Londrina in May 2006. The program evaluation was performed using the number of pregnant women and children treated at referral service and consumption of medicine for the treatment of gestational toxoplasmosis. The results revealed that there was a reduction of 63.9% and 42.6% in the number of pregnant women and children, treated at referral centers, respectively. In relation to the drugs, there was a 62.3% reduction in the consumption of folinic acid and 67.4% sulfadiazine. The standardization of diagnosis protocols, treatment and possible actions, brought to physicians, greater safety for decision making demonstrated by a decrease in pregnant women treated unnecessarily (Mitsuka-Breganó, 2009; Lopes-Mori et al., 2011). The actions of the program are included in the manual "Congenital Toxoplasmosis and Gestational: Manual for Health Surveillance, Diagnosis, Treatment and Conduct" (Mitsuka-Breganó et al, 2010). After a few years of the beginning of the program implementation and with the purpose to keep envolved professionals updated, technological resources are being developed to the stage of Technical Training Workshops in the implementation of the program in other cities and also to update the teams that are already in cities with the ongoing program, aiming and allowing interaction between the team and generating information on a wider scale and at great speed making the process effective.

# Conclusion

The results obtained through evaluation of the program in Londrina demonstrated a reduction in the number of pregnant women and children sent to referral services for the treatment of toxoplasmosis. Moreover, the definition of protocols results in the standardization of care and safety for decision making by the team of health care for pregnant women, thus reducing the consumption of drugs used in treatment (Mitsuka-Breganó, 2009).

The prevention actions of primary, secondary and tertiary of this model program of health surveillance of toxoplasmosis are strategies that can reduce the risks of exposure and prevent pregnancy and congenital toxoplasmosis.

The efficiency of a program that involves changes in lifestyle are associated with knowledge of the epidemiological aspects of the region, the standardization of service protocols, the establishment of a consensual ideal conduct and with technical support (Lopes-Mori et al., 2011), broad and repeated disclosure of actions highlighting the risk factors and the participation of all health professionals, pregnant women, family and community.

# References

Aspock, H.; Pollak, A. Prevention of prenatal toxoplasmosis by serological screening of pregnant women in Austria. *Scandinavian Journal of Infectious Diseases*, 1992, v. 84, Suppl., p. 3277.

Bachmeyer, C.; Mouchnino, G.; Thulliez, P.; Blum, L. Congenital toxoplasmosis from an HIV-infected woman as a result of reactivation. *The Journal of Infection*, London, 2006, v.52, n.2, p.e55–e57.

Bittencourt, Lhfb.; Lopes-Mori, F. M. R.; Mitsuka-Breganó, R.; Valentim-Zabott, M.; Freire R. L.; Pinto S. B.; Navarro It. Soroepidemiologia da toxoplasmose em gestantes a partir da implantação do Programa de Vigilância da Toxoplasmose Adquirida e Congênita em municípios da região oeste do Paraná. *Rev Bras Ginecol Obstet.* 2012; v.34, n.2, p. 63-68.

Boyer, Km.; Holfels, E.; Roizen, N.; Swisher, C.; Mack, D.; Remington, J.; Withers, S.; Meier, P.; McLeod, R. Risk factors for *Toxoplasma gondii* infection in mothers of infants with congenital toxoplasmosis: implications for prenatal management and screening. *American Journal of Obstetrics and Gynecology*, Saint Louis, 2005, v.192, n.2, p.564-571.

Cortina-Borja, M; Tan, H,K; Wallon, M; Paul, M; Prusa, A; et al. Prenatal Treatment for Serious Neurological Sequelae of Congenital Toxoplasmosis: An Observational Prospective Cohort Study. *PLoS Med,* 2010, v. 7, n. 10: e1000351. doi:10.1371/journal.pmed.1000351.

Dias, Rcf.; Lopes-Mori, Fmr.; Mitsuka-Breganó, R.; Dias, Raf.; Tokano, D.V.; Reiche, Emv.; Freire, Rl. & Navarro, It. - Factors associated to infection by *Toxoplasma gondii* in pregnant women attended in Basic Health Units in the city of Rolândia, Paraná, Brazil. *Rev. Inst. Med. Trop. São Paulo*, 2011, v. 53, n.2, 185-191.

Dressen, D.W. *Toxoplasma gondii.* J Am Vet Med Assoc, v.196, n.2, p.274-276, 1990.

Dubey, J. P.; Beattie, C. P. Toxoplasmosis of animals and man. Boca Raton: CRC, Press, p.1-220, 1988.

Dunn, D.; Wallon, M.; Peyron, F.; Petersen, E.; Peckhan, C.; Gilbert, R. Mother-to-child transmission of toxoplasmosis: risk estimates for clinical counseling. *Lancet*, 1999, v. 353, p. 1829-1833.

Elsheikha, H. M. Congenital toxoplasmosis: Priorities for further health promotion action. *Journal of the Royal Institute of Public Health,* 2008, v. 122, p. 335-338.

Escuissato, D. L.; Aguiar R. O.; Gasparetto, E. L.; Müller, N. L. Disseminated toxoplasmosis after bone marrow transplantation: high-resolution CT appearance. *Journal of Thoracic Imaging,* Philadelphia, 2004, v.19, n.3, p.207-209.

Foulon, W.; Naessens, A.; Derde, M. P. Evaluation of the possibilities for preventing congenital toxoplasmosis. *American Journal of Perinatology*, 1994, v. 11, p. 57-62.

Gilbert, R. E.; Peckham, C. S. Congenital toxoplasmosis in the United Kingdom: to screen or not to screen? *Journal of Medical Screening*, 2002, v. 9, p. 135–141.

Giraldi, N.; Vidotto, O.; Navarro, I. T.; Garcia, J. L.; Ogawa, L.; Kobylka, E. Toxoplasma antibody and stool parasites in public school children, Rolândia, Paraná, Brazil. *Revista da Sociedade Brasileira de Medicina Tropical*, Rio de Janeiro, 2002, v.35, n.3, p.215-219.

Hohlfeld, P.; Daffos, F.; Thulliez, P.; Aufrant, C.; Couvreur, J.; Macaleese, J.; Descombey, D.; Forestier, F. Fetal toxoplasmosis outcome of pregnancy and infant follow-up after in utero treatment. *Journal of Pediatrics*, 1989, v. 95, p. 11-20.

Hung, C.C.; Fan, C.K.; Su, K.E.; Sung, F.C.; Chiou, H.Y.; Gil. V.; Ferreira, M.D.A. C.; Carvalho, J.M.; Cruz, C.; Lin, Y.K.; Tseng, L. F.; Sao, K.Y.; Chang, W. C.; Lan; H.S.; Chou, S.H. Serological screening and toxoplasmosis exposure factors among pregnant women in the Democratic Republic of Sao Tome and Principe. *Transactions of the Royal Society of Tropical Medicine and Hygiene*, 2007, v. 101, n.2, p. 134-139.

Lago, E. G.; Neto, E. C.; Melamed, J.; Rucks, A. P.; Presotto, C.; Coelho, J. C.; Parise, C.; Vargas, P. R.; Goldbeck, A. S.; Fiori, R. M. Congenital toxoplasmosis: late pregnancy infections detected by neonatal screening and maternal serological testing at delivery. *Paediatric and Perinatal Epidemiology*, 2007, v. 21, p. 525-531.

Lappalainen, M.; Sintonen, H.; Koskiniemi, M.; Hedman, K.; Hiilesmaa, V.; Ammala, P.; et al. Cost-benefit analysis of screening for toxoplasmosis during pregnancy. *Scand J Infect Dis;* 1995, v. 27, p. 265-272.

Lebech, M.; Andersen, O.; Christensen, N. C.; Hertel, J.; Nielsen, H. E.; Peitersen, B.; Rechnitzer, C.; Larsen, S. O.; Nørgaard-Pedersen, B.; Petersen, E. Feasibility of neonatal screening for toxoplasma infection in the absence of prenatal treatment. *Danish Congenital Toxoplasmosis Study Group. Lancet*, 1999, v. 353, p. 1834-1837.

Logar, J.; Petrovec, M.; Novak-Antolic, Z.; Premru-Srsen, T.; Cizman, M.; Arnez, M.; Kraut, A. Prevention of congenital toxoplasmosis in Slovenia by serological screening of pregnant woman. *Scandinavian Journal of Infection Disease*, 2002, v. 34, p. 201-204.

Lopes-Mori, F. M. R. Epidemiologia da toxoplasmose gestacional em cinco municípios do estado do Paraná. Tese (Doutorado em Ciência Animal) – Universidade Estadual de Londrina, 2010.

Lopes-Mori, F. M. R.; Mitsuka-Breganó, R.; Capobiango, J. D.; Inoue, I. T.; Reiche, E. M. V.; Morimoto, H. K. et al. Programas de controle da toxoplasmose congênita. *Rev. Assoc. Med. Bras.*[online]. 2011, vol.57, n.5, pp. 594-599.

Mitsuka-Breganó, R. Programa de Vigilância em Saúde da Toxoplasmose Gestacional e Congênita: elaboração, implantação e avaliação no município de Londrina, Paraná. Tese (Doutorado em Ciência Animal) – Universidade Estadual de Londrina, 2009.

Mitsuka-Breganó, R; Lopes-Mori, F. M. R.; Navarro, I.T. Toxoplasmose Gestacional e Congênita: Manual de Vigilância em Saúde, Diagnóstico, Tratamento e Condutas. Editora EDUEL, 2010. P.66 – ISBN: 978-85-7216-519-8.

Montoya, J. G.; Liesenfeld, O. 2004. Toxoplasmosis. *Lancet*, London, v.363, n.9425, p.1965-1976, 2004.

Navarro ,I.T.; Vidotto,O.; Giraldi,N.; Freire, R.L. *Toxoplasma gondii*: Isolamento a partir de carne e cérebro de suínos comercializados na região de Londrina, Pr. Semina, 1992, v.13, p.15-18.

Neto, E. C.; Rubin, R.; Schulte J.; Giugliani, R. Newborn screening for congenital infectious diseases. Emerging Infectious Diseases, v. 10, p. 1068-1073, 2004.

Ortega-Benito, J. M. Prenatal screening of congenital toxoplasmosis. *Medicina Clinica, Barcelona*, 2001, v.116, n.10, p.385-389.

Paul, M.; Petersen, E.; Pawlowski, Z. S.; Szczapa, J. Neonatal screening for congenital toxoplasmosis in the Poznan region of Poland by analysis of Toxoplasma gondiispecific IgM antibodies eluted from filter paper blood spots. *The Pediatric Infectious Disease Journal*, 2000, v. 19, p. 30-36.

Pinkerton, H.; Weinman, D. Toxoplasma infection in man. Archives of Pathology, Chicago, 1940, v.30, p.374.

Rodrigues, T.O.; Perri, S.H.V.; Nunes, C.M.; Valladão, G.M.R.; Gallani, S.U.; Pinheiro, S.R.; Queiroz, L.H. Ações educativas para o controle de vetores da dengue e leishmaniose visceral. *Veterinária e Zootecnia* (UNESP), 2011, v. 18, p. 462-472.

Santos, A. N.; Mercado, L. P. L. Arquivamento e comunicação de imagens radiológicas na formação médica online. Rev. bras. educ. med. [online]. 2010, vol.34, n.4, pp. 525-534. ISSN 0100-5502.

Silva, L. A.; Vieira, R. S.; Serafini, L. N.; Carlotti Junior, C. G.; Figueiredo, J. F. C. Toxoplasmose do sistema nervoso central em paciente sem evidência de imunossupressão: relato de caso. Revista da Sociedade Brasileira de Medicina Tropical, Rio de Janeiro, 2001, v.34, n.5, p.487-490.

Silveira, C.; Belfort Jr., R.; Burnier JR., M. Identificação de cistos de *Toxoplasma gondii* na retina de irmãos não gêmeos com diagnóstico de toxoplasmose ocular recidivante: primeiro caso mundial. Arquivo Brasileiro de Oftalmologia, 1987, v.50, n.6, p.215-118.

Silveira, et al., 2001.

Tenter, A.M.; Heckeroth, A.R.; Weiss, L.M. Toxoplasma gondii: from animals to humans. *Int. J. Parasitol.*, 2000, v. 30, p. 1217-58.

Thulliez, P. Screening programme for congenital toxoplasmosis in France. Scandinavian *Journal of Infectious Diseases*, 1992, v. 84, Suppl., p. 43-45.

Wolf, A.; Cowen, D. Granulomatous encephalomyelitis due to an encephalitozoo (encephalitozic encephalomyelitis). A new protozoan disease of man. Bulletin of the Neurological Institute of New York, New York, 1937, v.6, p. 306-371.

Wolf, A.; Cowen, D.; Paige, B. Human toxoplasmosis; occurrence in infants as an encephalomyelitis, verification by transmission to animals, Science, Washington, 1939, v.89, p.226-227.

# Index

## C

## D



## E

## N



## O



## P

## T